THE QUANTUM SELF

THE QUANTUM SELF

Human Nature and Consciousness Defined by the New Physics

Danah Zohar

in collaboration with I. N. Marshall

QUILL / WILLIAM MORROW
NEW YORK

Grateful acknowledgment is made: to A. P. Watt, Ltd., on behalf of the Executors of the Estate of Robert Graves and Oxford University Press, New York, for permission to quote Robert Graves's "The Thieves"; to Faber and Faber, Ltd., and Harcourt Brace Jovanovich, Inc., for permission to quote from T. S. Eliot's "Four Quartets" and *Murder in the Cathedral;* to Ashford Publishing for permission to quote from A. E. Housman's "A Shropshire Lad"; to Basil Blackwell for permission to quote from Roger Penrose in Colin Blakemore and Susan Greenfield's *Mindwaves.*

Recognizing the importance of preserving what has been written, it is the policy of William Morrow and Company, Inc., and its imprints and affiliates to have the books it publishes printed on acid-free paper, and we exert our best efforts to that end.

Library of Congress Cataloging-in-Publication Data

Zohar, Danah, 1945-
The quantum self / by Danah Zohar in collaboration with I. N. Marshall.
p. cm.
Includes bibliographical references.
ISBN 0-688-10736-2
1. Quantum theory. 2. Physics—Philosophy. 3. Philosophical anthropology. I. Marshall, I. N. II. Title.
QC174.12.Z64 1990
530.1'2—dc20
89-13194
CIP

Printed in the United States of America

4 5 6 7 8 9 10

BOOK DESIGN BY PAUL CHEVANNES

For Anna Feodora and Ivan David,
without whom it would have been impossible;
with whom, it took a lot longer

Author's Note

Throughout this book I have drawn heavily on the thinking of my husband, Dr. I. N. Marshall—both that which he communicated to me in long, mutually creative discussions and that which he has published or is about to publish in academic or professional journals. The formal argument and mathematical rigor of the latter, though not generally accessible to the lay reader, lend added weight to the book's main thesis, and these publications appear in the Notes and Bibliography where relevant. Special note is made of his paper "Consciousness and Bose-Einstein Condensates."

Preface

This book had an odd beginning. Three years ago a television crew came to interview me about another book I had written, on precognition and modern physics.[1] I explained apologetically that it was difficult for me to think about anything so abstract just then because I was pregnant. When the producer asked we what I could talk about, I threw up my hands and answered, "Motherhood."

There followed, much to my surprise and theirs, a long conversation about motherhood and modern physics. I found myself describing my pregnant woman's psyche, the birth of my first child, and my sense of myself as a mother in terms of the often bizarre world of subatomic particles described in quantum physics. The strange picture of reality depicted in quantum theory seemed, at the very least, to offer a rich imagery for discussing the equally strange state of pregnancy and early motherhood. To my even greater surprise, this conversation became the basis of a television program on quantum physics and, eventually, part of a book.[2] It also reawakened something in me.

I first discovered quantum theory at the age of sixteen. I am certain that early introduction influenced both my life and my way of looking at the implications of what is generally called the new physics. In late adolescence so many things become uncertain and one is driven by a tremendous urgency to find answers to life's "big questions": who am I, why am I here, what is my place in the scheme of things, why is the world like it is, what does it mean that one day I must die? Where my parents' rather set-piece answers and the simple Methodism of my grandparents failed to offer any light, the new physics seemed to hold out to me a kind of poetic vision.

The seething equivalence of matter and energy, the flux suggested by the wave/particle duality, the sudden birth and death of particles that I witnessed in the vapor trails of my homemade cloud chamber, and the tantalizing unfixedness of reality suggested by Heisenberg's

Uncertainty Principle all worked like a kind of potion to excite my imagination, and gave me an admittedly somewhat mystical sense that the universe was "alive." My grasp of the mathematics of quantum theory was not sufficient at the time to garner any detailed explanation of the fundamental nature of things, but I had found the rudiments of a faith that "it all meant something."

Unfortunately, that was as far as my passion was to take me for some twenty years. Despite, or perhaps because of, an undergraduate degree in physics, I became caught up in other pursuits, in the business of getting on with life.

To most people, the world of physics seems a world apart. Its complex mathematical formulas, its apparently unfathomable experimental results, appear to bear no relation whatever to the everyday world of commonsense experience, no relation to our perceptions or emotions, never mind to the personal and social problems that occupy so much of our lives. And yet physics, like all science, began in the realm of daily experience. It began with wonder and with questions about how and why things worked, with the kinds of questions we all ask about the world and our place within it. And the answers to these questions affect us all, scientist and nonscientist alike.

Personally, I find myself very much preoccupied with them again—for instance, when talking to my husband (a psychiatrist and psychotherapist) about his work, about the structure of the brain and the vicissitudes of human consciousness, or when recalling the late Krishnamurti speaking about the interrelatedness of all things, proclaiming "I am the world."

Increasingly, since that original conversation with the television crew, I find myself drawing on my own knowledge of quantum physics. Its depiction of reality at the subatomic level and the really very strange goings-on among electrons have given me new insight into certain common philosophical problems, such as, personal identity (how much of me is really "me"; how much should "I" count?), the mind-body problem (how does my conscious mind, or "soul," relate to my material body or to other matter?), the problem of free will versus determinism, and the problem of meaning. Or again, when I am thinking about the experiences of daily life quantum physics has enriched my reflections about giving birth, my thoughts about dying, and my feelings of empathy or even telepathy between myself and others, and made me reconsider the way that the material world (for instance, very ugly inner cities) impinges on consciousness.

At times, quantum theory seems to serve as a useful metaphor that helps to draw these reflections into a new and sharper focus, while at others it seems to promise at least a partial explanation for how consciousness, and hence daily experience, might actually work. This book began primarily as an exercise in metaphor but, as it unfolded, metaphor gave way increasingly to evidence, or to what is at least well-grounded speculation about the actual physics of human psychology and its moral and spiritual implications.

In writing the book I was painfully aware that every chapter could—and in many ways should—have been a book in itself. But because the basic idea of seeing ourselves as quantum persons is itself so radically new, I felt it best as a first step to offer a broad overview, making it possible, I hope, for the reader to appreciate its far-reaching significance. Perhaps others will pursue some of its many themes in greater detail.

Many people have helped in making the book. I would like to thank especially the members of the Oxford Physics and Philosophy Group for hours of enlightening discussion, and the Oxford Psychotherapy Society for the uncanny relevance of its speaking program to themes I was trying to develop. The early faith and encouragement of my editor, Maria Guarnaschelli, and her sustained support throughout, gave me the confidence to write a much "bigger" book. My agent, Dinah Wiener, gave the book its start long before either she or I appreciated its scope.

I have already given (inadequate) mention to my husband's intellectual contribution. But beyond that, it was his patience, his constant good humor, and his untold hours of creative babysitting that ultimately made the book possible.

Also, I owe more gratitude than I can express to the City and the University of Oxford—for Port Meadow and her pubs, the "smells and bells" of St. Barnabas' Church on Sunday mornings, the countless beautiful buildings that lift the spirit at every turn, the libraries, lecture halls, and seminar rooms made open to all who need them, and the numerous dons so often available for conversation and suggestion. Quantum physics shows us that we cannot separate ourselves from our environment, and I doubt that I could have written this book living in any other place.

Contents

Preface *9*

Chapter 1 A Physics of Everyday Life *17*
Chapter 2 What's New About the New Physics? *24*
Chapter 3 Consciousness and the Cat *38*
Chapter 4 Are Electrons Conscious? *50*
Chapter 5 Consciousness and the Brain: Two Classical Models *62*
Chapter 6 A Quantum Mechanical Model of Consciousness *76*
Chapter 7 Mind and Body *92*
Chapter 8 The Person That I Am: Quantum Identity *107*
Chapter 9 The Relationships That I Am: Quantum Intimacy *125*
Chapter 10 The Survival of the Self: Quantum Immortality *141*
Chapter 11 Getting Beyond Narcissism: The Foundations of a New Quantum Psychology *154*
Chapter 12 The Free Self: Quantum Responsibility *171*
Chapter 13 The Creative Self: Ourselves as Coauthors of the World *188*
Chapter 14 Ourselves and the Material World: Quantum Aesthetics *203*
Chapter 15 The Quantum Vacuum and the God Within *216*
Chapter 16 The Quantum World View *231*

Notes *239*
Bibliography *247*
Index *257*

THE QUANTUM SELF

CHAPTER 1

A PHYSICS OF EVERYDAY LIFE

Several quite good popular accounts of quantum physics have been published in recent years. This book is not intended to be yet another. Rather than being about quantum physics per se, it is more a book about how the insights of modern physics can illuminate our understanding of everyday life, can help us better to understand our relationship to ourselves, to others, and to the world at large.

More specifically, it is a book whose central theme is about getting beyond a particular form of alienation that has plagued life in this century. This is the sense of alienation that follows from a feeling that we human beings are somehow strangers in the universe, merely accidental by-products of blind evolutionary forces, with no particular role to play in the scheme of things and no meaningful relationship to the inexorable forces that drive on the larger world of brute, insensate matter. To pursue this theme, I shall be looking very closely at the relationship between matter and consciousness in quantum theory and proposing a new, quantum mechanical theory of consciousness that promises to bring us back into partnership with the universe.

The roots of this alienation run deep in our culture, going back at least as far as Plato's philosophy with its distinction between the realm of Ideas and the world of experience, and later drawing on Christianity's denigration of the body in favor of the soul.

But by common consent, the strongest influences in our modern culture derive from the philosophical and scientific revolution of the seventeenth century, which encompassed the cultivation of Cartesian doubt and the birth of Newtonian, or classical, physics. Both changed radically the way we look at ourselves and our relation to the world. Cartesian philosophy wrenched human beings from their familiar social and religious context and thrust us headlong into what this book calls our I-centered culture, a culture dominated by egocentricity, by an overemphasis on "I" and "mine." Newton's vision tore us out from the fabric of the universe itself.

Classical physics transmuted the living cosmos of Greek and medieval times, a cosmos filled with purpose and intelligence and driven by the love of God for the benefit of man, into a dead, clockwork machine.

The Copernican revolution had already displaced the earth, and hence human beings, from the center of things, but Newton's three laws of motion and his mechanical model of the solar system were the blueprint for an entirely lifeless design. Things moved because they were fixed and determined; cold silence pervaded the once-teeming heavens. Human beings and their struggles, the whole of consciousness, and life itself were irrelevant to the workings of the vast universal machine.

Throughout history we have drawn our conception of ourselves and our place in the universe from the current physical theory of the day. Thus physicists and nonphysicists alike these three hundred years have found their personal philosophies, their own senses of identity, and their notions of how they relate to the world and to other people colored by this bleak Newtonian vision.

The immutable laws of history portrayed by Marx, Darwin's blind evolutionary struggle, and the tempestuous forces of Freud's dark psyche all, to some extent, owe their inspiration to Newtonian physical theory. All, together with the architecture of Le Corbusier and the whole vast array of technological paraphernalia that touches every aspect of our daily lives, have so deeply permeated our consciousness that we each see ourselves reflected in the mirror of Newtonian physics. We are steeped in what Bertrand Russell called the "unyielding despair" to which it has given rise.

"The world which science presents for our belief," Russell wrote at the turn of this century, tells us

> that man is the product of causes which had no prevision of the end they were achieving; that his origin, his growth, his hopes and fears, his loves and his beliefs, are but the outcome of accidental collocations of atoms; that no fire, no heroism, no intensity of thought and feeling, can preserve the individual life beyond the grave; that all the labours of the ages, all the devotion, all the inspiration, all the noonday brightness of human genius, are destined to extinction in the vast death of the solar system, and that the whole temple of Man's achievement must inevitably be buried beneath the debris of a universe in ruins. . . .[1]

"How," he asked, "in such an alien and inhuman world can so powerless a creature as man preserve his aspirations untarnished?" To a large extent we have not.

Most written accounts of our century, and the experience of a great many people who have lived through it, paint a picture of considerable dissolution. On every side—morally, spiritually, and aesthetically—our culture seems to be under stress. Many of the "old values" and generally held beliefs have ceased to be unquestionable. We find ourselves grounded in nothing larger than ourselves, and the great mass of people have been forced willy-nilly to live in the age of the existential hero—defiantly indifferent to the dead God, becoming makers of their own values and guardians of their own consciences. This is the experience of "modernism," and its cost in terms of both personal and cultural unrootedness has been high.

In our relationship both to ourselves and to others, the Newtonian influence runs deep. If we are nothing but accidental by-products of creation and pawns in the play of larger forces wholly beyond our control, how can we exercise much meaningful responsibility either for ourselves or towards others?

How, with our existence temporary and our purposes futile, tossed this way and that by the dynamics of the id or the undercurrents of genetic or class struggle and history, can we really be held accountable for anything? So much of modern sociology and educational theory, indeed our whole psychology of the person, follows from such thinking—as does our peculiarly twentieth-century violence, a natural reaction to so much impotence.

Equally affected is our attitude towards Nature and the material world. If our minds, or conscious selves, are wholly different from our material selves as Descartes argued, and if consciousness has no part to play in the universe as Newtonian physics implies, what relationship can we have to Nature and to matter? We are aliens in an alien world, set apart from and in opposition to our material environment. Thus we set out to conquer Nature, to overwhelm her and use her for our own ends, never minding the consequences.

"Man is a stranger to the world," says Michel Serres, "to the dawn, to the sky, to things. He hates them and fights them. His environment is a dangerous enemy to be fought, to be kept enslaved. . . ."[2] The twentieth-century desecration of the environment and the mindless proliferation of ugly, man-made material structures follow from this sense of alienation from Nature and from matter.

Yet ironically, while the "Newtonian world view" still dominates our lives and our thoughts, any excitement about Newtonian physics itself has long ago gone by the by. It is still the physics that drives dynamos and puts men on the moon,* but it is no longer at the forefront of creative physical thinking; it is not even taught to undergraduates at most front-ranking universities because it is considered suitable only for more elementary levels of science education. In its stead we have the "new physics," Einstein's relativity theory and quantum mechanics, both of which have changed radically the way that physics is done.

Relativity theory itself, while having important consequences for the way that some physics is done, is not likely to lead to a new world view. Though a misreading of Einstein has given some encouragement to the trend towards "relativism" in certain types of historical and anthropological thinking, relativity theory itself is about the physics of high velocities and very great distances. It plays itself out on a cosmological scale and has virtually no application in our everyday, earthbound world.

Thus, while every schoolchild knows that space is curved and time as we know it a "mere illusion," it is very unlikely that ordinary people will find their understanding of daily reality very much colored by Einstein's work.

---

*I often catch myself using *man, men,* and *he* for persons in general. No sexist bias is intended. I am simply from a generation that uses what may one day be called "the old English"!

Quantum physics is different. Being the physics of that tiny microworld within the atom, it describes the inner workings of everything we see and, at least physically, are.

The whole world of matter, including our own bodies, is made up of atoms and their even smaller components, and the laws that govern these tiny bits of basic reality spill over into our daily lives. A single photon, or "particle" of light, affects the sensitivity of the optic nerve. The Uncertainty Principle that rules the behavior of electrons plays a role in the buildup of genetic mistakes which contribute to the aging process and the development of certain cancers, and the process of evolution itself is thought to be similarly influenced.[3]

At the level of analogy, quantum physics is rich with imagery that almost begs for application to the experiences of daily life. Heisenberg's Uncertainty Principle long ago made its way into the language of sociologists and psychologists, the notion of a "quantum leap" has become common parlance for discussing any sort of rapid change, and, more amusingly, in the city of Chicago motorcycle repairmen have been seen sporting T-shirts with QUANTUM MECHANIC written across the front. In London, the Quantum Partnership is an advertising agency.

Throughout the course of this book I shall be drawing on a great many ways in which quantum theory can provide us with a radically new understanding of various aspects of our experience, and it is the overall theme of the book that a whole new metaphor for the age, or a new world view, follows naturally from what quantum physics is telling us about the physical and the human world. The characteristics of this world view will become clear as we discuss why the new physics is new and see how, through a new physics of consciousness, it can be applied to the philosophy of the person and the psychology of human relationships.

In some important ways, the subject of this book—how quantum physics relates to our experience of everyday life—goes straight to the heart of the central philosophical problem within quantum theory itself. So far, sixty years into their field's young history, quantum physicists still find themselves wholly unable to explain how there *can* be any everyday world—the world of tables and chairs, rocks and trees—let alone how their science might relate to it.

Quantum theory is our most successful physical theory ever. It can predict correct experimental results to an accuracy of several decimal points. But its inability to *explain* either the predictions or the results has meant that no one, new picture of reality itself has emerged from all the equations generated, never mind a new world view in which the discoveries of quantum physics reach down to quicken the imaginations of ordinary people.

Indeed, for most of the past sixty years since quantum theory was completed, the dominant view among quantum physicists has been that they neither could nor should say anything about the real world, that their only "safe" task was to stick to predicting results through their equations.

This "antirealist" view, which has come to be known as the Copenhagen Interpretation of quantum theory after the Danish physicist Niels Bohr, who advocated it most strongly, is influenced by the bizarre and shadowy nature of quantum-level events, where nothing in particular can be said to exist in any fixed place and everything is awash in a sea of possibilities. It has led to some absurd talk among quantum theorists and their philosophical followers, including a denial that there *is* any reality at the subatomic level, or even, in some cases, a denial that there is any reality *at all.*

But there is a real world in which "things" exist. Chairs are solid and our identifiable, solid bodies can sit on them. Unlike Schrödinger's pet, which I shall discuss later, cats are either alive or dead, and when my son kicks his ball it lands either in our own garden or in the neighbor's. For quantum theory really to come into its own, and to replace not just Newtonian physics but also the whole Newtonian world view as the central philosophy of our time, it must be brought more into dialogue with such facts in the everyday world.

It is the central argument of this book that we conscious human beings are the natural bridge between the everyday world and the world of quantum physics, and that a closer look at the nature and role of consciousness in the scheme of things will lead both to a deeper philosophical understanding of the everyday and to a more complete picture of quantum theory.

The existence of consciousness has always been a problem. What is it, why is it here in the world at all, how can such a thing *be?* Some answer to these questions is necessary to any understanding of life even in its most primary sense, such as the "aliveness" of an amoeba. In a broader sense still, some answer is needed to illuminate the meaning and purpose of life, the whys and wherefores of our culture, and the place of a single individual in the larger universe. It is also necessary to any understanding of the universe itself.

In this book I shall be considering very seriously the possibility that consciousness, like matter, emerges from the world of quantum events; that the two, though wholly different from each other, have a common "mother" in quantum reality. If so, our thought patterns—and beyond that, our relationships to ourselves, to others, and to the world at large—might in some ways be explained by, and in other ways mirror, the same laws and behavior patterns that govern the world of electrons and photons.

If our intellect does indeed draw its laws from Nature, then we have the further consequence that our perception of these laws must to some degree mirror the reality of Nature herself. Thus in knowing ourselves, we can come to know Nature.

If such a possibility exists, then, as Michel Serres suggests, we can draw from it a vision, similar to that of the ancient Greeks:

> . . . Where man is in the world, of the world, in matter, of matter, he is not a stranger, but a friend, a member of the family, and an equal. . . . The Greeks lived in a reconciled universe. Where the science of things and the science of man coincide.[4]

It is my conviction that in quantum physics we now have the foundations of a physics upon which we can base both our science and our psychology, and that through a wedding of physics and psychology we, too, can live in a reconciled universe, a universe in which we and our culture are fully, and meaningfully, part of the scheme of things.

CHAPTER 2

WHAT'S NEW ABOUT THE NEW PHYSICS?

Einstein once said that quantum theory reminded him of "the system of delusions of an exceedingly intelligent paranoic, concocted of incoherent elements of thought."[1] All the descriptions commonly applied to this physics—*absurd, bizarre, mind-boggling, incredible, beyond belief*—are of one sort. Even finding the really most apposite means of describing discoveries in the field seems an elusive task.

The new physics is so new that quantum physicists themselves haven't fully come to terms with the conceptual changes it demands, taking refuge instead in the less demanding language of mathematics. Yet it is here, in forging a new conceptual structure for the new physics, that the real cultural challenge of modern science lies.

Old intellectual habits die hard. The Newtonian categories of space, time, matter, and causality are so deeply ingrained in our whole perception of reality that they color every aspect of the way that we think about life, and we can't easily imagine a world that mocks their reality.

Each time we drive a car from one point to another, we are to some extent conscious of the space between the two points and the time it takes to travel between them. The simple act of opening and closing a door makes us subliminally aware of the material being of both the door and our hand, and of the cause-and-effect relation between them.

How, then, do we deal with a claim that there is no space between separate objects, indeed that there are no objects such as we normally think of them and that the whole notion of "separate" has no foundation in reality? How do we talk about events or relationships if we must give up all talk of time and never say that one thing causes another to happen?

Such problems, when first presented, create a kind of intellectual numbness, followed then by an attempt to deal with them in some familiar way. Even quantum physicists, when they try to make sense of what their equations are telling them, unwittingly try to force the new quantum concepts into old Newtonian categories, in turn usually finding their own work as bizarre as it seems to the outsider. So far, not a single one has succeeded in saying what it all *means.*

Throughout this book I shall be trying to express the concepts of quantum theory in ordinary language and in terms of the everyday. Hopefully I shall avoid the usual trap of trying to fit square pegs into round holes. The radical newness of it all will be at once apparent as we look at the basic notions of being, movement, and relationship in the new physics. I feel confident that our ability to assimilate it as part and parcel of our own experience will grow in later chapters.

BEING

The most revolutionary, and for our purposes the most important, statement that quantum physics makes about the nature of matter, and perhaps being itself, follows from its description of the wave/particle duality—the assertion that all being at the subatomic level can be described equally well either as solid particles, like so many minute billiard balls, or as waves, like undulations of the surface of the sea. Further, quantum physics goes on to tell us that neither description is really accurate on its own, that both the wavelike and the particlelike aspects of being must be considered when trying to understand the nature of things, and that it is the duality itself that is most basic. Quantum "stuff" is, essentially, *both* wavelike and particlelike, simultaneously.

This Janus-like nature of quantum being is summed up in one of the most fundamental tenets of quantum theory, the Principle of Complementarity, which states that each way of describing being, as a wave or as a particle, complements the other and that a whole picture emerges only from the package deal. Like the right and left hemispheres of the brain, each description supplies a kind of information that the other lacks. Whether at any given time elementary being displays itself as one or the other depends on the overall conditions—crucial among which, as we shall see later, may be whether or not anybody is looking, or what the observer is looking for! "Elementary particles," said Sir William Bragg, "seem to be waves on Mondays, Wednesdays and Fridays, and particles on Tuesdays, Thursdays and Saturdays."[2]

Such duality and the somewhat ethereal concept of matter couldn't be further from the more everyday notion held in Newtonian, or classical, physics.

In Newton's physics, to which we owe our common perception of larger things, it was assumed that being, at its most basic, unanalyzable level, consisted of tiny, discrete particles—atoms—which bump into, attract, or repel each other. These particles were solid and separate, each occupying its own definite place in space and time. Wave motions, such as light waves, on the other hand, were thought to be vibrations in some underlying "jelly" (the ether), not fundamental things in themselves. Thus both waves and particles had a role to play in Newtonian physics, but particles were thought to be more basic, and it was these of which matter consisted.

For quantum physics, however, both waves and particles are equally fundamental. Each is a way that matter can manifest itself, and both together are what matter *is.* And while neither "state" is complete in itself, and both are necessary to give us a complete picture of reality, it turns out that we can never focus on both at once. This is the nub of Heisenberg's Uncertainty Principle, the other most fundamental principle of being in quantum theory.

According to the Uncertainty Principle, the wave and particle descriptions of being preclude one another. While *both* are necessary to get a full grasp of what being is, only *one* is available at any given time. Either we can measure the exact position of something like an electron when it manifests itself as a particle, or we can measure its momentum (its speed) when it expresses itself as a wave, but we can never measure both, exactly, at the same time.

The measuring conundrum for electrons is a bit like the dynamics of a first psychiatric interview, in which, ideally, the psychiatrist would like to know *both* the relevant background facts about his patient *and* establish some sort of rapport with him. The trouble is, if the psychiatrist asks factual questions to elicit the history, he gets simply factual answers, while the patient himself, his way of being at that moment, fades into the background. On the other hand, if the psychiatrist decides to abandon questions for more creative, receptive listening, he will get a good feel for the patient, but conclude the interview knowing very little about his history. Fact gathering and rapport seem to exclude each other, and yet each is necessary for a total picture of the patient's condition.

Similarly, most electrons and other subatomic entities are *neither* fully particles *nor* fully waves, but rather some confused mixture of the two known as a "wave packet," and this is where the wave/particle duality and quantum mystery come fully into their own. While we can measure wave properties, or particle properties, the exact properties of the *duality* must always elude any measurement we might hope to make. The most we can hope to know about any given wave packet is a fuzzy reading of its position and an equally fuzzy reading of its momentum.

This essential fuzziness is the uncertainty referred to in the name of the Uncertainty Principle, and it replaces the old Newtonian determinism, where everything about physical reality was fixed, determined, and measurable, with a vast "porridge" of being where nothing is fixed or fully measurable. Here everything remains indeterminate, somewhat ghostly, and just beyond our grasp.

Just as we often feel we can never fully understand other persons, never really pin them down in their essence, it is certainly true that we can never fully know an elementary particle. It's as though we were forever condemned to seeing only shadows in the fog. The full nature of this quantum indeterminism goes straight to the heart of the central philosophical problem raised by quantum mechanics—the nature of reality itself.

Some quantum theorists, foremost among them Niels Bohr and Heisenberg himself, argued that fundamental reality itself is essentially indeterminate, that there is no clear, fixed, underlying "something" to our daily existence that can ever be known. Everything about reality is and remains a matter of probabilities. An electron *might* be a particle, it *might* be a wave, it *might* be in this orbit, it *might* be in that—indeed, *anything* might happen. We can only predict such things on the basis of what is most probable given the overall constraints of any given experimental situation.

On this view, where the essential basis of reality as we know it consists of just so many possibilities, we are left with the central unanswered problem of quantum theory being: How can anything in this world *ever* become actual or fixed? It's the very opposite of the dilemma raised by Newton's clockwork universe, in which there is no scope for the new. Reading Newton, we have to ask: How can anything ever *happen?* With the Bohr-Heisenberg interpretation of quantum mechanics, the great problem becomes: How can anything ever *be?*

But other quantum theorists, led by an impassioned Einstein, have argued that any such thoroughly indeterminate, probabilistic reality is beyond conception. The Almighty, Einstein assured us, would not countenance a universe run according to the rules of a common gaming house—"I shall never believe that God plays dice with the world," he proclaimed.

Instead, the Einstein camp has argued that the essential indeterminism required by quantum mechanics does not lie in reality itself, but follows rather from an incompleteness within quantum theory itself or from an inability on our part to study Nature without disturbing her. They point out that the theory fails at the point where it has to account for the existence of actual things, and insist that the actual world is as fixed and real as we ever thought it to be. It is we, because of the measurement processes or the equations we employ, who cannot know it.

While agreeing with Einstein that quantum physics as presently constructed fails to give an adequate account of the everyday material world we see around us, my own bias is towards the Bohr-Heisenberg view of indeterminancy. That is, the view that the foundation of reality itself is an unfixed, indeterminate maze of probabilities. The reason for this bias will become clearer when we come to discuss the nature of consciousness and its relation to quantum mechanics in later chapters. The functioning of our own minds may provide a key to the nature of fundamental reality.

For now, quantum indeterminancy is, at the very least, a powerful metaphorical way of perceiving reality. At the level of the everyday, we can see the Uncertainty Principle and the Principle of Complementarity—the wave/particle duality—as offering us a choice between different ways of looking at the same system. For instance, we can think of waves as massive undulations on the surface of the sea, or we can think of them as so many individual disturbed water "particles" (molecules). We can think of a nation as a living entity with its own characteristics, ethos, and history, or we can break it down into individual cities, buildings, and people.

If we take it still further, we can think of the bricks in the buildings or the body cells in the people, or even the molecular or atomic makeup of each. Different kinds of things can be seen more clearly from different perspectives, and who is to say which is more fundamental? which, or what, more actually exists?

Quantum field theory takes us even further beyond Newton's dead and silent universe, giving us a vivid picture of the dynamic flux that lies at the heart of an indeterminate being. Here, even those particles that do manifest themselves as individual beings do so only briefly.

At high enough energies, particles can be born out of a background of pure energy (waves), exist for the briefest while, and then dissolve again into other particles or return to the background sea of energy—like the short-lived vapor trails one sees in a simple Wilson cloud chamber, which emerge apparently from nowhere, traverse a brief space in the mist, and then disappear once again. Some of the transient, individual particles' properties—their masses, charges, and spins—are conserved, but the number and types of particles are not constant. As in the rise and fall of a nation's population or the construction and decline of its individual cities or buildings, such constancy is reserved for the overall balance of the entire system.

This graphic picture of the emergence and return, or the beginning and ceasing, of individual subatomic particles at the quantum level of reality holds deep implications for our way of looking at the nature and function of individual personalities, or the survival of the individual self.

MOVEMENT

In classical physics, movement seems a simple enough concept, familiar to us in our everyday perception of the way that things get about. An object, say a ball, travels continuously from point A to point B, takes a given amount of time to make its transition from one to the other, and makes its journey in the first place because someone threw it. Thus it moves smoothly through space and time as a result of cause and effect. We all know that this is the basic way events in our world are constructed.

Yet at the quantum level of reality, the whole picture of continuous movement through space and time breaks down. Quantum physics is, as one Oxford physicist puts it, a physics of "lumps" and "jumps."[3]

The lumps appeared in the early days of quantum theory when Max Planck proved that all energy is radiated in individual packets, called quanta, rather than in flowing streams over a continuous spectrum; the jumps appeared a few years later when Niels Bohr demonstrated that electrons jump from one energy state to another in discontinuous "quantum leaps," the size of the leap depending on how many quanta of energy they have absorbed or given off.

The original Bohr atom, now somewhat outdated but still useful for picturing the effect of quantum leaps, was like a minute solar system. It had a comparatively large nucleus at the center in the place of the sun and the various electrons circling about it in their individual orbits, each orbit representing a given energy state that an electron can occupy. There was, it turned out, no rhyme or reason as to when an electron might leap from one orbit to another, or how big a leap it might make. All that could be predicted with certainty was that its path would be bumpy and that the "distance" (energy difference) it traveled could be measured in so many whole quanta.

The new description of motion as a series of disrupted leaps was one of the most fundamental conceptual changes to come out of quantum theory. It was like replacing the smooth flow of real life with jerky, broken stills that make up the individual frames of a film. Indeed, the theory showed us that all motion—even that which we perceive as smooth and continuous—is structured like the successive presentation of frames of a film. And just as a film might occasionally skip a bit in the projector, so subatomic particles can leap several "frames" ahead, leaving out the intermediate steps that might seem more natural. The analogies with mental and cultural processes are legion.

As we have already seen in the discussion of being in quantum theory, Heisenberg's Uncertainty Principle arose from the problem of trying to follow and describe the actual movement of a subatomic particle along its discontinuous path. In a realm where reality appears to consist not of any fixed actualities that we *can* know but rather of the probabilities of all the various actualities that we *might* know, the harder one tries to scrutinize the movements of any given particle, the more elusive it becomes. The elusiveness is one major problem raised by quantum movement; the other is the fate of all those lost probabilities.

If reality at the everyday level on which we commonly experience it does indeed consist of actual things like bodies and desks and chairs, while at the quantum level there exist no actual "things" but rather myriad *possibilities* for countless actualities, what becomes of all that potential? At what stage, and why, does one of Nature's manifold possibilities fix itself in the world of "real things," and what role, if any, is played by all the lost possibilities in achieving this final state of affairs? The answer to these questions will interest us later on when we come to discuss the nature and function of consciousness.

As we shall see in the next chapter, there is as yet no good answer to the "why" of actuality, but the somewhat startling role of possibility in fixing, or even creating, it is better understood.

We have seen that when an electron makes a transition from one energy state to another within the atom, it does so in a completely random and spontaneous way. Suddenly, with no prior warning and certainly without "cause," a previously quiet atom may experience chaos in its electron energy shells. It's all just a matter of chance. And the electrons may, with equal probability, make a transition from a higher-energy state to a lower one, or from a lower-energy state to a higher one. This is why there is said to be time reversibility at the quantum level. Things can happen in any direction.

There is no familiar succession of events within the disturbed atom, with one thing causing another. Things just happen as they happen, as the loosely connected images within a poem may follow one upon the other in no particularly necessary order. Worse still, which brings us to the question of the lost possibilities, *they happen simultaneously in every direction at once.*

When an electron, in the guise of a probability wave, intends to move from one orbit to another, it at first behaves as though "it were smeared out over a large region of space,"[4] displaying a kind of eerie omnipresence in many orbits. It puts out temporary "feelers" towards its own future stability by way of trying out—all at once—all the possible new orbits into which it might eventually settle, in much the same way as we might try out a new idea by throwing out imaginary scenarios depicting its many possible consequences.

In quantum theory, these temporary "feelers" are called virtual transitions, whereas the electron's final transition into a new, permanent home is called a real transition. However, as quantum physicist David Bohm cautions, we shouldn't be misled by the terms *virtual* and *real*.

> Sometimes permanent (i.e., energy conserving) transitions are called *real* transitions, to distinguish them from the so-called *virtual* transitions, which do not conserve energy and which must therefore reverse before they have gone too far. This terminology is unfortunate, because it implies that virtual transitions have no real effects. On the contrary, they are often of the greatest importance, for a great many physical processes are the result of these so-called *virtual* transitions.[5]

The situation is a bit like that of a sheltered young woman at last presented to society at her coming out. Previously very much at rest, she finds herself quite overexcited when confronted by offers of marriage from several suitors. A whole new world of possibilities has opened to her and she naturally wants to realize her greatest potential for a happy marriage to the man of her dreams. In the real world (the world of everyday reality) she would have to explore these possibilities one by one, perhaps dating each of her suitors several times before feeling certain she could settle on just the right one.

But in the quantum world, the dizzy girl would simply take up with *all* suitors, *all at once,* perhaps even setting up house with each of them simultaneously. If her scandalized parents wished to write to her to reprimand her for this profligate behavior, they would find it impossible to pin her down. They would just have to send duplicate letters to all of her new addresses, since she really is to be found at all of them. And if the girl's love nests were close enough to one another, she could even stand on her own various front balconies and wave at herself!*

*In fact, for quantum theory the houses could be any distance apart, because an electron's virtual transitions interfere with each other over an infinite distance.

In the end, of course, having explored her possibilities to the full, the girl would eventually settle down, marry, and live in one house with just one of the suitors, but not without having left traces of herself in the various neighborhoods where she had occupied temporary addresses. The neighbors might remember seeing her and ask themselves what had become of her, and if nature had followed its usual course there might be offspring of the many temporary liaisons, who in turn would grow up to influence the world. (". . . For a great many physical processes are the result of these so-called *virtual* transitions.")

While the case of the quantum-level hussy might seem somewhat farfetched, there is a major interpretation of quantum theory that seriously argues that this sort of actualized multiple choice really happens every time there is a point of decision about which way an indeterminate physical process might resolve itself. Called the Many-Worlds Theory, it suggests that there are an infinite number of worlds, in each one of which we could find a version of our own selves, each different from the others in that it had pursued and developed some other possible chain of events. According to this view, there are no lost possibilities—we can have them all.

I shall not be pursuing the literal many-worlds interpretation in later chapters, though some later discussions will draw out analogies between psychological processes and the role of quantum virtual transitions. In biology, David Bohm has already suggested that "in many ways, the concept of a virtual transition resembles the idea of evolution in biology, which states that all kinds of species can appear as a result of mutations, but that only certain species can survive indefinitely, namely, those satisfying certain requirements for survival in the specific environment surrounding the species."[6]

The many species thrown up by mutations can be seen as various possibilities (virtual states) being explored by Nature as new ways through which she might express her potential. The less viable possibilities do, as Bohm says, eventually die out, but often not without leaving some traces of themselves, which go on to become part of life's fabric. Two unviable mutations might, for instance, crossbreed to form some third species that is capable of long-term survival (a real transition). Quote possibly we human beings are the result of such a crossbreeding between two "virtual species," a successful secondary mutation of some shadowy life form known only as the missing link.

RELATIONSHIP

Perhaps more than anything else, quantum physics promises to transform our notions of relationship. Both the concept of being as an indeterminate wave/particle dualism and a concept of movement that rests on virtual transitions presage a revolution in our perception of how things relate. Things and events once conceived of as separate, parted in both space and time, are seen by the quantum theorist as so integrally linked that their bond mocks the reality of both space and time. They behave, instead, as multiple aspects of some larger whole, their "individual" existences deriving both their definition and their meaning from that whole.

The new quantum mechanical notion of relationship follows as a direct consequence of the wave/particle dualism and the tendency of a "matter wave" (or "probability wave") to behave as though it were smeared out all over space and time. For if all potential "things" stretch out infinitely in all directions, how does one speak of any distance between them, or conceive of any separateness? All things and all moments touch each other at every point. The oneness of the overall system is paramount. It follows from this that the once ghostly notion of "action-at-a-distance," where one body can influence another instantaneously despite there being no apparent exchange of force or energy, is, for the quantum physicist, a fact of everyday reality. It is a fact so alien to the whole framework of space and time that it remains one of the greatest conceptual challenges raised by quantum theory.

A vision of reality that holds truck with instantaneous action-at-a-distance, or nonlocality as it is more properly called (the principle that something can be affected in the absence of a local cause), has obvious mystical overtones. It flies directly in the face of both common sense and classical physics. Both rest on the intuitive principle that reality is composed, at some level, of basic, unanalyzable parts. These parts are inherently separate, and any witnessed effect on one part is commonly assumed to have an attributable cause in some other part. Further, according to relativity theory, no cause (say, a signal) can travel from one bit of reality to affect any other bit faster than the speed of light. Thus any notion of instantaneous influences should be out of the question. The whole problem of nonlocality is so difficult that it was seldom even raised in the early days of quantum theory, and it's only in recent years that physicists have attempted to come to terms with it.

It was Einstein who first demonstrated that the equations of quantum theory predicted the necessity of instantaneous nonlocality. For him, it was impossible ("ghostly and absurd," he called it) and he never was comfortable with the wider metaphysical implications of quantum physics. The prediction of nonlocality was the clear proof he needed that quantum theory was "incomplete and wrong-headed," and it remained a strong plank in his campaign to get this recognized. In one of the famous paradoxes of physics—the Einstein, Podolsky, Rosen (E.P.R.) Paradox—he demonstrated, once and for all he thought, that the supposed existence of nonlocal influences led to a contradiction.

The gist of the E.P.R. Paradox can be understood by imagining the fate of a hypothetical set of identical twins separated since birth.* While one twin still lives in London, the other has gone off to live in California. Over the years there is no contact between the twins; indeed, they are ignorant of each other's existence. In the commonsense view of things, the twins have led entirely separate lives. Yet despite their separation and the lack of communication between them, a psychologist studying the twins has noted an amazing correlation in their life-styles. Each twin has adopted the nickname of Badger, each works as a prosecutor in the office of a district attorney, each dresses almost exclusively in shades of brown, and each married a blonde named Jane at the age of twenty-four. How can all this be explained?

The quantum physicist would have no difficulty in believing the twins' correlated lives. He would say his equations always predicted it and that all links between the twins are sufficiently explained by their individual existences' being aspects of some larger whole. But Einstein thought that was not enough. In his Theory of Hidden Variables, he suggested instead (staying with the analogy of the twins†) that there must be some common factor, say their shared genetic material, which

---

*The actual E.P.R. Paradox concerns a thought experiment proposed by Einstein, Boris Podolsky, and Nathan Rosen in which a physicist might try to measure the positions and momenta of two protons as they flew off in opposite directions from a common source. David Bohm later revised this by suggesting physicists measure the spins on two protons, and his suggestion became the basis for real-correlation experiments done in the seventies with photons, or "particles" of light.

†The example of the twins is my own, not Einstein's.

predetermined the similarity in their life-styles. The controversy was eventually settled by a physicist named John Bell, whose theorem led to definitive experiments.

Following the gist of Bell's Theorem, which calls for interfering with one member of a pair to see what happens to the other, we would at some point give the twin living in London a good kick down some stairs so as to break his leg. No one would argue that any shared genetic material could explain it if the twin living in California were then to suffer a similar fall. So if the California twin remained standing upright when his London counterpart was kicked, quantum theory would be proved wrong and Einstein correct; but if the California twin fell, Einstein would be wrong and quantum theory correct. In fact, when the London twin is kicked, the California twin should suffer an identical fall at exactly the same moment and break his leg, too, although no one has kicked him. All aspects of their lives are inseparable.

At the subatomic level, such correlation experiments have now been carried out many times on pairs of correlated photons, and the nonlocal influences that bind their "life-styles" have been proved many times over. The photons' behavior patterns are so linked across any spatial separation—it could be a few centimeters, it could be all the way across the universe—that it appears there is no space between them. Similar experiments have been done to show the same eerie correlation effects across time. Two events happening at different times influence each other in such a way that they appear to be happening at the same time. In fact, they manage to reach across time in some synchronized dance that defies all our common-sense-bound imagination.[7]

Imagine, for example, that there are two boatmen who work to ferry goods across a river, each with his own boat. Boatman A ferries goods with one boat, Boatman B with another. When traffic on the river is very busy, the two boatmen work simultaneously, but during slack periods they decide to work in shifts. Boatman A works mornings, and Boatman B afternoons.

During the busy spells when the boatmen work at the same time, they choose quite arbitrarily which boat to use, so neither regards either boat as his own. When they begin working in separate shifts, this randomness in selecting which boat to use persists—but with a crucial twist. When Boatman A arrives for the morning shift, he arbitrarily selects one of the boats to use; when B arrives for his afternoon shift, he always uses whichever boat A has not used (though he has no way of knowing which boat A has used). Thus, though the two boatmen are arriving for work at different times of the day, they continue to use the two boats in a way that suggests both are present. Their behavior is linked across the time gap between their shifts so that it is always correlated.

The correlations demonstrated in a photon experiment done along these lines were always so exactly symmetrical that it makes no sense to say that Boatman A chose one particular boat in anticipation of B's choosing the other, or that B chose his because of some mysterious knowledge of which one A had chosen earlier. All that can be said is that the correlations show how two events can be related across time in a way that ensures they will always act "in tune," and any attempt to set up a cause-and-effect relationship between them is useless. Such synchrony is the basis of all quantum mechanical relationships, lending a very modern note of support to the pre-Socratic Greek notion of the "oneness of being."

The extent to which correlated nonlocal influences exist between apparently separate bodies or events depends on the extent to which a system is in a "particle" or a "wave" state. Particles behave more like separate individuals and are less correlated; waves display a greater grouplike correlated behavior pattern. I shall return to this in later chapters when discussing personal identity and the roots of alienation.

The existence of nonlocal quantum-level correlations has shaken the world of physics and is one of the main factors making it so far impossible for quantum physicists to say what their theory means. Throughout this book I shall be drawing on analogies between quantum nonlocality and experiences in our everyday lives and relationships. But in Chapter 5, where I discuss in detail the nature of consciousness, nonlocal correlations between apparently separate bodies will be crucial to the discussion of consciousness as a quantum mechanical phenomenon. At that point it will become important to ask whether the new concept of relationship underpinned by nonlocality doesn't afford us a key to an entirely new understanding of ourselves.

CHAPTER 3

CONSCIOUSNESS AND THE CAT

> Seven years we have lived quietly
> Succeeded in avoiding notice
> Living and partly living
>
> —T. S. Eliot
> *Murder in the Cathedral*

Those who have read any of the many popular books on quantum mechanics will have encountered Schrödinger's cat. Like the women of Eliot's chorus, his lot is one of living and partly living. The poor beast suffers from a peculiarly quantum identity crisis, being indefinitely suspended in an intangible state where he is neither alive nor dead. His unhappy plight has generated more speculation and controversy than almost any other problem raised by the new physics, not least because it raises the question of human consciousness and its possible role in the formation of physical reality. In many ways it is the real starting point of many of the themes to be developed later on in this book.

It was clear in the last chapter that the central conundrum facing quantum physics, and those who would use it to talk about the world, is not "How can anything happen?" but rather "How can anything *be*?" If, as mainstream quantum theorists believe, reality at its most fundamental level is just an indeterminate porridge of many possibilities, a teeming flux of hybrid matter waves, how do we ever get the familiar world of definite, solid objects that we see around us? At what point, and why, is reality real-ized? To illustrate the problem and its paradox, Erwin Schrödinger, one of the founding fathers of quantum theory, introduced his cat into the debate.

Schrödinger's cat has been placed in one of those ubiquitous laboratory cages used for animal experimentation, only in this case the walls of the cage are solid. This is crucial, because to understand the point of the paradox, we mustn't see the cat until the end of the story.

Inside the opaque cage, Schrödinger has devised a macabre experiment. He places in the cage a bit of radioactive material that has, to keep things simple, a 50 percent chance of shooting out a decay particle in an upwards direction and a 50 percent chance of shooting one out in a downwards direction. If the decay particle shoots upwards, it strikes a particle detector, which in turn triggers a switch that releases lethal poison into the cat's feeding dish. The cat eats it and dies.

Similarly, if the decay particle shoots downwards, a switch is triggered that releases food, and the cat lives to fight another experiment.

That choice of outcomes, at least—"up," he dies; "down," he lives—is the choice we would expect in the everyday world. But things are not so simple for quantum cats. Indeed they are not simple at all, because according to mainstream quantum theory, the cat is *both* alive *and* dead. He exists in a superimposed state of both conditions at once, just as electrons are said to be *both* waves *and* particles at the same time (Figure 3.1).

Just like the quantum hussy who was able to live with all of her lovers simultaneously, the being of Schrödinger's quantum mechanical cat is "spread out" through space and time. His possible aliveness and possible deadness "fan out" as a probability wave to fill the cage. The best we can do to pin him down is to describe *all* of his possible states with a Schrödinger wave function—that is, with a mathematical equation that lists his many possibilities, just as the rules of poker lay out the many possible hands we might draw and what we can do with them while leaving us ignorant of exactly which hand we will actually be dealt. That is a matter of probability.

In this case, the wave function ("the rules") tells us that the cat has eaten the poison and died (Possibility I) *and* that the cat has enjoyed a nourishing meal and lives (Possibility II). It is only when this wave function "collapses," when all the possibilities it describes suddenly jell into one fixed reality, that we get a cat we can either bury or fondle.

Figure 3.1. The live/dead quantum cat

Obviously such a collapse must occur sometime because, so the story goes, when we open the cage and look at him, the cat is definitely dead (Figure 3.2).* But why? What killed Schrödinger's cat?

This question, which applies not just to quantum mechanical cats but to ourselves and everything we see around us, goes straight to the heart of why there is any reality at all, and it illustrates why the cat's identity crisis raises a paradox.

It is a paradox because, on the one hand, the world is full of quite ordinary cats that are either dead or alive, while on the other hand, the physics that has occupied the best scientific minds of our century tells us this is impossible. The mathematics of Schrödinger's equation argues that nothing *can* settle the cat's fate—nothing can collapse his wave function. At least nothing physical. Any physical object put into his cage—a camera, for example—that could tell us whether he is alive or dead would be struck with the Midas touch of too much possibility.

*Or alive. At any rate, his fate has been resolved.

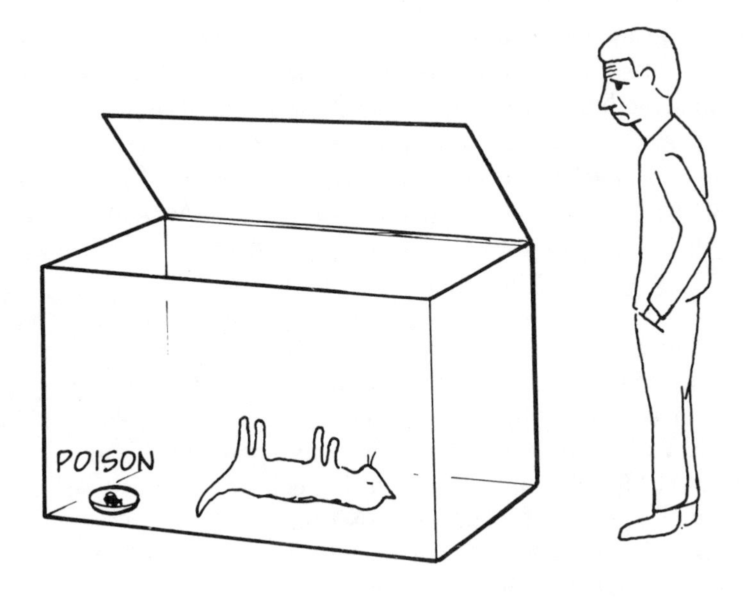

Figure 3.2. The cat is dead.

It would exhibit classic quantum mechanical behavior and become all things to all people.

So, despite the evidence of our own eyes, quantum theory says the cat is, and must always remain, both dead and alive. Not surprisingly, this paradox has been dubbed the observation problem, both because it challenges our commonsense observations and because it highlights the intriguing role of observation (and the observer) in molding reality.

REALITY HAPPENS WHEN WE LOOK AT IT

From its earliest days, quantum theory has implied that something very odd, and of crucial importance, happens when we observe a quantum system. Unobserved quantum phenomena are radically different from observed ones—that is one major point of the story surrounding Schrödinger's cat. At the moment of observation, or measurement, previously unobserved electrons that were both waves *and* particles become waves *or* particles, unseen single photons that had in some mysterious way managed to travel through two slits at the same time suddenly choose to go through one or the other, and mixed-up cats become something we can relate to. In short, something about that moment when an indefinite, many-possibilitied quantum wave function is seen (or recorded) collapses it into a single and fixed actuality. Schrödinger's cat wasn't simply *found* dead when we looked at him. In some strange way that nobody yet understands, he died precisely *because* we looked at him. Observation killed the cat.

That much is quantum fact—something about the act of observation (or measurement) collapses the quantum wave function*—and that fact alone is pregnant with implications we shall explore later. But because it is a fact without explanation, indeed a fact that shouldn't *be,* it leaves all the interesting questions unanswered and has led, understandably enough, to a great deal of quantum speculation[1]—and to no small amount of quantum confusion.

While we are naturally curious to understand why, after all, looks *can* kill, there is little point in getting ourselves lost in all this confusion. Solving the collapse of the wave function is well beyond the scope of this chapter. But, as one of the more common speculations about why it happens might easily be confused with the overall thesis of this book, I feel it important to distance myself from that speculation early on.

I want to argue that there is a physics of consciousness and that this physics suggests a great deal to us about the link between ourselves and physical reality. The *basis* of my argument about the link, however, is very different from the view put forward by those who usually claim that it was consciousness itself that killed Schrödinger's cat. Their use of consciousness as an effective feline eradicator rests on an understanding of the nature of consciousness wholly different from that which I shall suggest later.

A minority of physicists (and a majority of their popularizers) have proposed that, as quantum theory clearly shows us that nothing physical could have killed the cat, there must be some nonphysical explanation for his death. Some deus ex machina, as it were, enters the situation from outside the laws of physics to rescue Schrödinger, his cat, and all the rest of us from too much possibility. This metaphysical

---

*More accurately, it is at least *one* of the things that has this effect on quantum systems. There may be other, still unknown things that can collapse the wave function.

reality agent can't be the observer's measuring apparatus, or his eyes, or his brain, which are all physical and thus all covered by Schrödinger's equation. Therefore, they conclude, it must be the observer himself who kills the cat—that is, the observer's disembodied, immaterial consciousness.

According to this view, proposed chiefly by quantum physicists John Archibald Wheeler and Eugene Wigner, human consciousness is the crucial missing link between the bizarre world of electrons and everyday reality. Such a conclusion, ironically, is very close to my own, but my reasons for reaching it are radically different, and the difference is important to everything that follows later in this book.

Those who conclude that consciousness collapses the wave function because its nature is essentially nonphysical commit themselves and quantum physics to the old Cartesian view that mind and matter are separate entities. They view consciousness as something necessarily outside the physical world and thus as something alien to it—"the ghost in the machine." They also leave the door open to antirealist speculations that "reality is all in the mind," that there can't *be* any world unless somebody is looking at it,[2] and leave us wondering how we all got here in the first place. What conscious being was here at the beginning of things to collapse the first wave function?

My own reasons for suggesting that consciousness is an important link between the quantum and everyday worlds have a very different starting point. The whole project of defining a new "quantum self" rests on arguing that quantum physics, and more particularly a quantum mechanical model of consciousness, allows us to see ourselves—our souls, if you like—as full partners in the processes of Nature, both "in matter" and "of matter."* This sort of argument has very different implications for someone who is trying to understand how we conscious creatures relate to everything else in the universe.

My own guess, were I to speculate at all about the death of Schrödinger's cat, would be in sympathy with that of the physicists who suggest there is nothing paradoxical about reality itself, but rather something wrong—or at least incomplete—about quantum theory. In its present form, since it can't account for whatever it is about observa-

*It will become clear later (in Chapters 6 and 7) that this quantum view of consciousness does not commit us to the familiar reductionist argument that the mind is nothing but a collation of atoms.

tion that collapses the wave function, the theory simply can't apply to the whole of physical reality. We must need some further mathematics, they argue—perhaps even the discovery of wholly new physical principles*[3]—before we can understand the transition from the quantum world to our own. Given that mathematics or those principles, everything will literally fall into place. Personally, I believe that what we need—at least in part—is a better physics for observers themselves, for their consciousness.

In distancing the argument of this book from the notion that the observer's disembodied mind killed the cat, I am not, therefore, denying that human consciousness has a creative role to play in the formation of physical reality. Indeed many of the themes developed in later chapters rest on claiming that it does, and at the level of everyday reality the evidence is almost too obvious to mention. Every time a conscious person chooses to lift his arm, consciousness is having an effect on physical reality. A hewer of wood or a builder of buildings has a still more creative effect.

But the ability of consciousness to affect quantum processes goes much further, striking, apparently, at the very heart of reality formation and raising tantalizing questions about the nature of both consciousness and reality.

HOW REALITY HAPPENS DEPENDS ON HOW WE LOOK AT IT

We have already seen that the act of observing quantum systems changes them into ordinary objects. The mere fact of our interference in Nature transforms her, and that fact alone would require that we change our whole way of looking at ourselves and our place in the natural world. But worse still for those who like to think that the world "just is as it is," our interference has an unexpected dimension.

Not only does observation somehow collapse the wave function, thus

---

*Principles possibly to do with the interaction between wave functions and the physical systems (wave functions collapse when they interact with other, larger physical systems—such as the measuring apparatus or even the observer's brain) or with gravity (wave functions collapse when they get sufficiently heavy).

helping to give us a world in the first place, but it turns out that the particular *way* in which we choose to observe quantum reality partly determines what we shall see. The quantum wave function contains many possibilities, and it can be up to us which of these will be elicited.

A photon, for example, has both position possibilities (a particlelike nature) and momentum possibilities (a wavelike nature). A physicist can set up his experiment to measure, and hence fix, either of these—though in fixing one he loses the other (Heisenberg's Uncertainty Principle). His interference—his measurement or observation—seems in some strange way to influence which side of its nature the photon will exhibit. The thought experiment about Schrödinger's cat isn't complex enough to illustrate this, but another experiment conceived by Wheeler does so graphically.*[4]

If a photon is given the option to travel through either one or both slits in a screen (being quantum mechanical it has the option to do both), the physicist's experiment will have the following result. If he places two particle detectors to the right of the slits, he finds that the photon behaves like a single particle—it follows a definite path through one slit and strikes one particle detector (Figure 3.3).

If, on the other hand, he places a detector screen between the two slits and the particle detectors, the photon behaves like a wave—it travels through *both* slits, interferes with itself, and leaves an interference pattern on the detector screen (Figure 3.4).

Physicist and photon are involved in a creative dialogue that somehow transmutes one of many quantum possibilities into an everyday, fixed reality. Therefore, the act of measurement does play some role in deciding what gets measured. "In some strange sense," says Wheeler, "this is a participatory universe."[5]

> Beyond particles, beyond fields of force, beyond geometry, beyond space and time themselves, is the ultimate constituent [of all there is], the still more ethereal act of observer-participancy?[6]

To capture the flavor of this observer-participancy, Wheeler recounts an old Hebrew legend. In the legend, Jehovah and Abraham are

---

*The following description is a very simplified version of the Wheeler delayed-choice experiment.

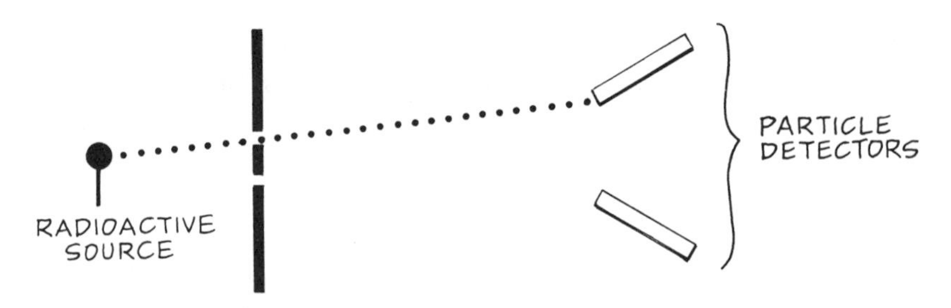

Figure 3.3. If you observe the photon with a particle detector, you get a particle.

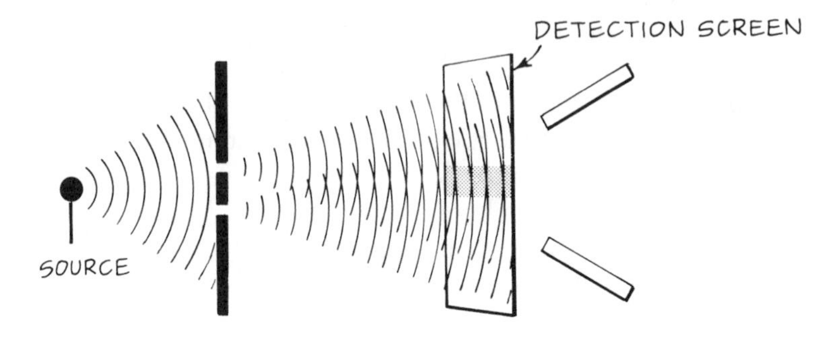

Figure 3.4. If you observe the photon with a "wave detector" (a screen on which an interference pattern can be seen), you get a wave.

having a heated dialogue about who has the upper hand in accounting for why the world is as it is.

"You would not even exist if it were not for me," Jehovah reminds Abraham. "Yes, Lord, that I know," Abraham replies, "but also you would not be known if it were not for me."[7] In more scientific language, Nobel laureate Ilya Prigogine makes the same point when he says, "Whatever we call reality, it is revealed to us only through an active construction in which we participate."[8]

In quantum physics, this interdependency between the *being* of a thing and its overall environment is called contextualism, and the implications of it are vast, both for our whole conception of reality and for our understanding of ourselves as partners in that reality. It is one central reason for my claim that quantum theory must contribute eventually to a new world view, with its own distinctive epistemological, moral, and spiritual dimensions. The epistemological dimension—what is the nature of our knowledge and what do we mean by truth?—was expressed very well in the phenomenology of French philosopher Maurice Merleau-Ponty, in what he called "truth within a situation":

> So long as I keep before me the ideal of an absolute observer, of knowledge in the absence of any viewpoint, I can only see my situation as being a source of error. But once I have acknowledged that through it I am geared to all actions and all knowledge that are meaningful to me, then my contact with the social in the finitude of my situation is revealed to me as the starting point of all truth, including that of science, and, since we have some idea of truth, since we are inside truth and cannot get outside it, all that I can do is define a truth within the situation.[9]

I shall say more about this, and about the moral and spiritual dimensions of observer-participancy, in later chapters, but a word of caution about quantum contextualism—"truth within a situation"—is necessary here.

Misunderstood and pushed in the wrong directions, the fact that the human observer in some way helps to evoke the reality that he observes could have unfortunate cultural implications. It could lend the full weight of physics to the currently popular, and in my view very pernicious, notion that the individual self is the sole author of value—

that there is no "truth" in this world but only one's "perspective."[10]

To some extent, certain of the popular books already written about quantum physics have encouraged their readers to draw such conclusions. Consider, for example, the epistemological and moral implications of Fritjof Capra's claim that, as "the mind of the observer creates the properties possessed by electrons," those properties can in no sense be called objective. Concerning atomic physics he says:

> In transcending the Cartesian division between mind and matter, modern physics has not only invalidated the classical ideal of an objective description of nature but has also challenged the myth of a value-free science. . . . The scientific results [scientists] obtain and the technological applications they investigate will be conditioned by their frame of mind.[11]

Mainstream quantum theory itself carries within it the dangers of such subjectivism (to wit, Heisenberg: "The conception of objective reality has thus evaporated . . ."[12]), but Capra pushes it further by introducing the notions of value and frame of mind. Such thinking is dangerous, and what is more, it is bad physics.

Nothing in quantum theory itself suggests that observation or the observer "creates" reality (the properties of subatomic particles). At the moment of observation, some dialogue between the quantum wave function and the observer (be this man or machine) *evokes,* and thus gives concrete form to, one of the many possible realities inherent within that wave function. But there is already the potential for some very definite sort of reality there—the wave function of a table can't collapse into a cat or a kangaroo. It can become only a table.

Furthermore, once the wave function has collapsed, its reality is as objective as anything else science studies. Any two (or more) people looking at Schrödinger's cat will agree that he is objectively dead—he won't look dead to one and alive to another. His mortality is not a matter of anyone's "point of view," and certainly not of someone's "value judgment." He is just simply, and finally, dead.

The whole large set of questions raised by the conundrum of Schrödinger's cat, among them the role of the human observer in reality formation and the associated problem of objectivity, only highlight the fact that at this stage we haven't enough understanding of human observers and the physics of their consciousness to reach any informed conclusions. The problem of the cat obviously obliges us to rethink a great many of our preconceptions about ourselves and possibly about the purpose of our existence, but to meet this heady challenge, we must face head-on the problem of consciousness.

CHAPTER 4

ARE ELECTRONS CONSCIOUS?

The problem of Schrödinger's cat raises the conundrum of the conscious observer's participation in reality formation and suggests that this may be an issue for physics itself. But this, in turn, raises further problems, problems that affect our attitude toward biology, psychology, philosophy, and religion—the whole vast range of disciplines that have as their goal any understanding of human beings and our place in the universe. Physics, today, is at the center of our concerns, and the problem of consciousness within physics is one of the most central of all.

But, while what the observer sees can be described in the equations of quantum mechanics, the observer himself cannot. We don't have any equations for observers, human or otherwise. They are outside the quantum system. Thus, ironically, while urging us to transcend the old observer/observed duality, quantum physics as presently understood actually sustains it. It is somehow awkwardly incomplete and will remain so until it can take account of observers and, in the case of human ones at least, the consciousness with which they make their observations.

But the consciousness that has become an issue for physics may be more than just human consciousness. When considering the plight of Schrödinger's cat, why shouldn't we wonder how his odd predicament seems to the cat himself or how it is affected by his consciousness? Or by that of a flea in his ear? Or, outrageous as the question may seem at first, by that of the radioactive decay particle that determines whether he lives or dies?

Something broader than the issue of the individual alone, or of the individual's relationship to matter, may be at stake. Something about the behavior of fundamental reality as expressed in the new physics almost demands that we reassess the whole question of consciousness, not just as it relates to ourselves but also as it may relate to other creatures and things in the universe—perhaps, as we shall see, even to the most elementary constituents of matter.

On the whole, the Judeo-Christian tradition, which informs much of our Western cultural history and self-awareness, has set man apart as something unique, certainly in this world and possibly in the universe as a whole. According to this tradition, God made all the creatures of the field each after its own kind, but He made man in His own image and gave him dominion over all the earth.[1] Man owed his special position not to his body, which consisted of mere "clay," but to his possession of a soul—in modern terms, his consciousness—which in some way mirrored that of the Divine Being. In modern philosophical terms all this was clarified and passed on to us in Descartes's mind/ body dualism, in the splitting of reality into thinking substances (*res cogitans*) and purely mechanical, extended substances (*res extensa*).

Given faith in such a transcendent deity, it matters less that man's soul, or consciousness, bears little relation to anything else in this world. If we are at one with God, what need have we of communion with the beasts and inanimate things? "My kingdom is not of this world."[2]

But with the advent of modern science in the seventeenth century and the slow but relentless withdrawal of the transcendent deity from the scheme of things, our human consciousness seemed no longer to mirror anything but itself. Without the Christian God, without faith in a transcendental kingdom of the soul and blind to the "soulfulness" (consciousness) of creatures and things, atheistic Cartesian dualism left us with nothing but crude materialism.* A sense of being unique because chosen gave way to the more familiar twentieth-century sense of alienation: We are distinct from everything around us and inexorably

*Like Newton, Descartes himself, of course, was a believing Christian, so for him Cartesian dualism was not the spiritual problem that it became for later generations.

alone. For some time it was even the vogue among modern psychologists and philosophers—the behaviorists and the positivists and linguistic analysts—to respond to this alienated uniqueness by denying its *raison d'être,* by denying the importance of consciousness altogether and the relevance of the whole world of subjective thought and emotion.

> The time seems to have come when psychology must discard all reference to consciousness; when it need no longer delude itself into thinking that it is making mental states the object of observation. . . . Psychology as the behaviourist views it is a purely objective, experimental branch of natural science which needs introspection as little as do the sciences of chemistry and physics.[3]

Ironically, this kind of thinking is now as outdated for physics as it was crippling for psychology.

The Cartesian world view was necessary to the cultivation of Newton's physics and all the technological progress that flowered in its wake, but in a post-Christian culture it is both philosophically and spiritually barren. While the soul of modern man cries out for more, for some sense of fellow feeling with something beyond ourselves, for a sense of being at home in the universe, our reason, too, demands that we make better sense of our experience. Consciousness is a fact of that experience, and a philosophy or a science that can't account for consciousness is a necessarily incomplete philosophy or science. This has become almost a home truth to quantum physicists struggling to make some sense of developments in their own field, but this truth has yet to percolate into our general intellectual outlook.

What are the implications, then, if both Christianity and prequantum modern science have got it wrong? What if man is not unique? What if, after all, we share to some extent our being conscious with other things and creatures in this universe—perhaps with the universe itself? Such questions become impossible to ignore if we take into account the knowledge of modern biology, or take seriously suggestions by philosophers and physicists like Alfred North Whitehead and David Bohm that even elementary subatomic particles might possess rudimentary conscious properties.

Before we go on in later chapters to explore the nature of human consciousness, its possible physics, and the psychological and moral implications of this, it would be useful here to reappraise the question of where conscious beings as a whole fit into the general scheme of things. What can we say now about this "consciousness" that we keep referring to, and who else—or what else—has it? Are we human beings really distinct from everything else, as the mainstream Western tradition has held, or is our consciousness in some important sense continuous with other things in the universe? And if continuous, how far does this continuity extend? To dogs and cats? To amoebas? To stones? Or even to electrons? In even beginning to think like this, we are experiencing something of a paradigm shift.

OTHER LIVING THINGS

Only the most extreme advocates of human uniqueness would deny the conscious life of familiar mammals like dogs and cats. They are clearly not anesthetized (insensible)—the most bare-bones criterion of something's being conscious. They move about, engaging in spontaneous and purposive activity; they have an undoubted capacity to enjoy pleasure and pain; they learn from their environment and adapt to it; and they are at least to some extent in possession of free will—they can, and do, make choices. All of these things we commonsensically associate with consciousness in human beings. Whether dogs and cats also enjoy, as we do, an "inner life" or possess a sense of "I" is a moot point with advocates on both sides, but we generally have no difficulty in feeling they are fellow conscious beings.

As we move further away from the common mammals encountered in daily life or further down the phylogenetic scale, the sense of fellow feeling becomes less gripping. Arguments from analogy—we are conscious, therefore things that are like us must be conscious—lose their force with the appearance of increasingly stranger creatures that are not very much like us at all. This is one of the problems raised by the philosopher Thomas Nagel in his much-discussed essay "What Is It Like to Be a Bat?"[4] When a creature's whole sensory experience and life-style are so different from our own, it becomes difficult for us to know "what it is like to be" that creature, i.e., what sort of inner life or experience it might have.* Nonetheless, if we thought about it, most of us would ascribe some sort of conscious life to bats or ants or possibly even to

*Nagel's famous criterion for consciousness: "But fundamentally an organism has conscious mental states if and only if there is something that it is like to *be* that organism—something it is like *for* the organism." (Nagel, 1979, p. 166.)

earthworms, and biologists, whose experience of these things is wider than most, are willing to go further still, easily seeing such organisms as amoebas and sea anemones as fellow conscious creatures.

"The writer is thoroughly convinced," says H. S. Jennings in W. H. Thorpe's classic study of animal behavior, "after a long study of the behaviour of *Amoeba*, that if *Amoeba* were a large animal, so as to come within the everyday experience of human beings, its behaviour would call forth to it states of pleasure and pain, of hunger and desire, and the like, on precisely the same basis as we attribute these things to the dog."[5] Thorpe himself goes on to say that "the behaviour of even a sea anemone is vastly more complicated than supposed. Not only is there a great deal of spontaneous movement but there are elaborate patterns of apparently purposive activity,"[6] all of which, he points out, would be easily available to any of us if we viewed the creature's activities on speeded-up film.

Further evidence that lower animals that are outwardly quite different from ourselves nonetheless possess some sort of conscious awareness that operates on principles similar to our own was suggested recently by studies of the effects of common surgical anesthetics (such as halothane and chloroform) on the behavior of great pond snails.[7] When exposed to the same anesthetics that deprive human patients of their conscious awareness, the snails were found to lose their withdrawal reflex in the presence of pain stimuli.

For our part, it is probably safe, even on present evidence, to assume that when we speak of consciousness we speak of a "property" or a process that we human beings share, at least to some degree, with all other members of the animal kingdom. This assumption embraces the intuitive feelings that many of us have towards other animals, and it accepts the possible validity of philosophical arguments from analogy.

Thus, in varying degrees of quality and complexity, we can grant that all other animals are in some sense aware, capable to some degree of spontaneous and purposive activity, sensitive to stimuli something like pleasure and pain, and in possession of some rudimentary capacity to exercise free will.* In the most primitive sense possible, possession of

*To list the possession of free will as a necessary criterion of something's being conscious is of course controversial, but it will be argued in Chapter 12 that consciousness—viewed as a quantum mechanical process—and free will are indeed inseparable. A quantum self is necessarily a free self.

this set of qualities would also imply some sort of subjective "inner life" on the part of other animals—every creature must have its own "point of view." Acceptance of this might well affect our moral stance towards creatures other than ourselves.

PANPSYCHISM—STRONG AND LIMITED

Most people would probably have little difficulty accepting the thesis so far, that is, accepting at least the possibility that all members of the animal kingdom possess a conscious life to some extent. Some of us might need convincing that snails have a "point of view" or that earthworms possess free will, but it is not wholly outside our scope to imagine that other creatures might share some of the properties we normally associate with conscious awareness.

A smaller number are at least familiar with the notion—if not wholeheartedly convinced of it—that other living things, such as plants, might also be endowed with some sort of primitive sentient properties. But if we go beyond that to a panpsychist position, suggesting that even inanimate objects like stones or logs (never mind electrons) are to be counted among nature's conscious beings, we go well beyond the intuitions of most people—at least those influenced by the intellectual climate of the last three hundred years. Very few of us alive today have any sense of fellow feeling with the dirt we tread upon or the dust we breathe in.

Yet our modern intuition about such things is at odds with many strands of our own pre-Cartesian, pre-Newtonian cultural history. Some form of organized panpsychism has been with us since pre-Socratic times. The One of Parmenides and the divine Flux of Heraclitus imply that all things, conscious and material, derive ultimately from one common source. "God is day and night, winter and summer, war and peace, surfeit and hunger; but he takes various shapes, just as fire, when it is mingled with spices, is named according to the savour of each. . . . Men do not know how what is at variance agrees with itself."[8]

Earlier still, the Nature spirits of the animists inhabited the trees and mountains and thunderclouds of ancient Greece, as they did in many other primitive societies. The metaphor of the Great Chain of Being, which portrays everything as belonging to one unified and complete chain extending from man to the smallest particles of inanimate matter, originated in Plato's *Timaeus* and influenced the world view of people throughout the medieval and Renaissance periods.[9] It is only in the modern era that we have largely lost touch with this ancient paradigm.

But despite, or more likely as a defensive reaction against, the materialistic and mechanistic bent of our recent culture, panpsychism in one form or another has developed a subcultural modern tradition of its own. For many the motivation has been primarily spiritual or religious. As the *Encyclopedia of Philosophy* expresses it, many have believed that "it is only by accepting panpsychism that a modern man (who finds it impossible to believe in the claims of traditional religion) can escape the distressing implications of materialism."[10]

By raising matter to the level of consciousness, or at least by seeing some nascent conscious properties in all matter, many modern philosophers and psychologists (Spinoza, Leibnitz, William James, Teilhard de Chardin, Whitehead, and so on) have made contact with an underlying reality not wholly alien to their own experience.

"If we are panpsychists," wrote the German philosopher Rudolf Hermann Lotze in the eighteenth century, "we no longer 'look on one part of the cosmos as but a blind and lifeless instrument for the ends of another,' but, on the contrary, find 'beneath the unruffled surface of matter, behind the rigid and regular repetitions of its working . . . the warmth of a hidden mental activity.' "[11] Lotze's contemporary G. T. Fechner saw the earth itself as a living creature, "a unitary whole in form and substance, in purpose and effect . . . and self-sufficient in its individuality"[12]—an idea made current in our own time by the enthusiasm felt for J. E. Lovelock's *Gaia* hypothesis.[13]

Many of the early modern panpsychists accepted the doctrine in its fullest form, believing that every mountain, tree, flower, and dust particle actually possesses an inner psychological life, but this is not the sort of panpsychist thinking that need concern us here.

Our concern is to see what light modern physics can cast on the nature of consciousness, to understand what it is about the relationship between matter and consciousness at the quantum level that now causes some quantum physicists, and a handful of philosophers informed by their work, to be counted among those in the panpsychist tradition. Theirs is of necessity a much more cautious or limited form of panpsychism, as there is nothing whatever about modern physics to suggest that mountains have souls or that dust particles possess an inner life.

The logic employed in limited panpsychism begins with a set of obvious facts. There is only one basic kind of matter, all things—animate and inanimate—are made of it, some of this matter has the undoubted capacity for conscious life, and at the quantum level at least there is a creative dialogue between matter and consciousness. This dialogue means that the observer's conscious mind actually influences the material development of that which he observes. As the philosopher Thomas Nagel expresses it:

> Each of us is composed of matter that has a largely inanimate history before finding its way onto our plates or those of our parents. It was once probably part of the sun, but matter from another galaxy would do as well.* . . . Anything whatever, if broken down far enough and rearranged, could be incorporated into a living organism. No constituents besides matter are needed.[14]

Furthermore, the inanimate matter that we conscious beings are made of keeps changing—in the case of human beings, it changes *totally* every seven years. Not one single atom now contributing to the makeup of my physical being was any part of me seven years ago. Our living bodies are in constant, dynamic interchange with other bodies and with the inanimate world around us. So how can the very same atoms be part of a conscious structure at one point in their history and of an inanimate object at another? At what point do they, or the structure of which they are a part, acquire consciousness? In his essay on panpsychism, Nagel arrives reluctantly at the conclusion that

> unless we are prepared to accept . . . that the appearance of mental properties in complex systems has no causal explanation at all, we must take the current epistemological emergence of the mental as a reason to believe that the constituents have properties of which we are not aware, and which do necessitate these results.[15]

---

*"In a sense, human flesh is made out of stardust. Every atom in the human body, excluding only the primordial hydrogen atoms, was fashioned in stars that formed, grew old and exploded most violently before the Sun and Earth came into being." (Nigel Calder, quoted in Carol Hill, 1986, p. 210.)

That is, we must accept that unless consciousness is something that just suddenly emerges, just gets added on with no apparent cause, then it was there in some form all along as a basic property of the constituents of all matter. As Karl Popper says, "Dead matter seems to have more potentialities than merely to produce dead matter."*[16]

But when Nagel suggests that some aspects of mind, or consciousness, might be associated with all matter, he is speaking of what he calls "proto-mental properties"—some primitive mental aspect of reality that only properly becomes conscious when suitably combined in a complex system. He argues that both these proto-mental properties and the elementary matter with which they are associated might derive from a common source, from a more fundamental level of reality that itself has a two-sided potential to become both the mental and the material.

> Such reducibility to a common base would have the advantage of explaining how there could be necessary causal connections in either direction, between mental and physical phenomena.[17]

Nagel's description of a more fundamental reality that is the common source of both the mental and the material aspects of the world is certainly compatible with what is known about quantum reality and the wave/particle duality. His position is shared by some leading quantum physicists. David Bohm, for instance, informed by his long career in physics and clearly influenced by the panpsychist thinking of Spinoza and Whitehead, believes that

> the mental and the material are two sides of one overall process that are (like form and content) separated only in thought and not in actuality. Rather, there is one energy that is the basis of all reality. . . . There is never any real division between mental and material sides at any stage of the overall process.[18]

For Bohm, as for Whitehead and Teilhard de Chardin before him, this process view of reality leads him to consider the presence of proto-

---

*Popper himself, however, is not a panpsychist. Unlike Nagel, he does believe that consciousness is an emergent phenomenon, a property of higher, complex systems but not of atoms.

conscious (Nagel's proto-mental) properties at the level of particle physics.

We saw in the last chapter that in some strange way an electron or a photon (or any other elementary particle) seems to "know" about changes in its environment and appears to respond accordingly. At least this is true under experimental conditions, and it is one of the more mysterious spin-offs of the observation problem.

In the famous two-slit experiment used to illustrate the wave/particle duality, photons behave quite differently depending on whether, before detection, they are offered the chance to pass through one slit in a screen or through two. If only one slit is open, they behave like particles, hitting the detecting surface like a stream of so many bullets. If two slits are open they behave like waves, passing through both slits and creating a typical interference pattern on the other side (Figure 4.1). They seem to "know" which aspect of their double-sided nature is called for by the experiment and behave accordingly.

In the Wheeler delayed-choice photon experiment discussed in the last chapter, this "knowledge" of the experimental setup is truly uncanny. In that experiment both slits are open at all times. But it is only later in its path that the photon encounters either a particle detector or an interference screen, one of which has been placed in its way *after* it has already passed through one slit or both. Even at that late stage, the photon seems to "know" what lies ahead and seems almost retroactively to choose both its own flight path and hence its nature (see Figures 3.3 and 3.4, page 46). It's only *after* it strikes one obstacle or the other that we can say whether it went through one slit or both.

Bohm uses a beautiful and evocative analogy to illustrate the apparent "knowing" properties of subatomic particles. He compares the movements of electrons in the laboratory to those of ballet dancers responding to a musical score, the score itself constituting "a common 'pool' of information that guides each of the dancers as they take their steps. . . .

> In the case of the electrons, the "score" is of course the wave function.* As with the dancers, the electrons are thus *participating* in a

---

*The wave function is that "listing" of all the electron's possibilities as described in the Schrödinger wave equation.

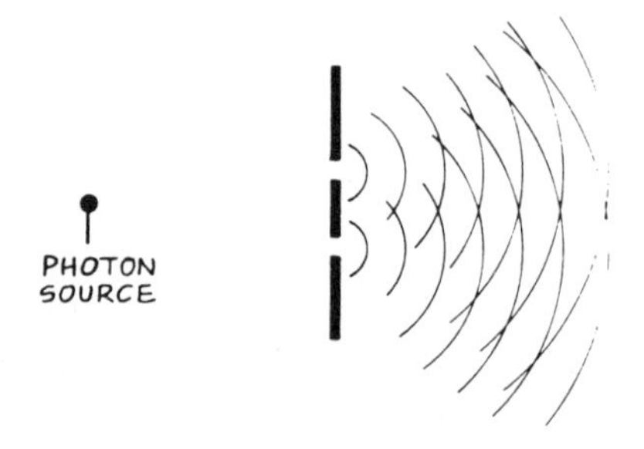

Figure 4.1. The photon seems to "know" how many slits are open.

common action based on a pool of information, rather than pushing or pulling on each other mechanically according to laws like those of classical physics.[19]

Each electron is sensitive not just to the information, or meaning, latent in its own wave packet (its own part in the score). It is also nonlocally responsive (owing to quantum correlation effects) to the information latent in the whole situation—the movements of other electrons, the design of the experimental apparatus, perhaps even the physicist's conscious intentions. For Bohm, this sharing of information, this mutual "knowing," may represent elementary conscious awareness on the part of the electron.* If he is correct—or, less boldly, even if there is something about quantum events that now makes it meaningful to raise such a possibility—this becomes yet another way in which the new physics is pushing us to shift our whole way of looking at the material world and our relation to it.

---

*Or, at least, the electron's participation in the dual mental/physical nature of meaning and information.

But we must be cautious. To say that a limited panpsychist view is compatible with quantum physics is not to say that it is necessitated by it. There is nothing in quantum theory as it has been developed so far that has anything whatever to say about the origins of consciousness in quantum reality, or about there being possible proto-conscious properties associated with elementary subatomic particles. Even Bohm's electron dance is at this stage just an engaging metaphor. Such possibilities are suggested by the uncanny behavior of photons and electrons in the laboratory and the participatory nature of the observer/observed relationship, but quantum theory per se has yet to take them on board—and indeed it has no way to do so until we achieve a better understanding of the nature of consciousness itself.

Ultimately, any really useful discussion of the possible conscious properties of elementary particles, or, indeed, of the relationship of matter and consciousness at all, calls for a wedding of physics and psychology that can be made possible only by a good model of how consciousness actually works—a model of the physics of consciousness. Such a model might then be used to explore the question of whether the familiar consciousness that we recognize as such in ourselves and other animals is an emergent property of complex living systems, or whether living systems simply have the capacity to organize the proto-conscious properties of more elementary matter in a meaningful way.

CHAPTER 5

CONSCIOUSNESS AND THE BRAIN: TWO CLASSICAL MODELS

While consciousness is in many ways the most familiar and accessible thing that each of us possesses, it remains one of the least understood phenomena in this world.

Every time we say "I" or "we," we draw on a tacit assumption that there *is* a conscious "I" or "we" doing the talking or thinking. But the moment we try to focus on this thinking self, to grab hold of it in some tangible way—as we might grab hold of a finger or an ear—it seems to vanish before our reach. We know a great deal about how fingers grasp and how ears hear, but about the origin and nature of that conscious person who originates the grasping or interprets the hearing, we have virtually no physical facts. There is no anatomy or physiology of consciousness, never mind a physics.

There are those, of course—the dualists—who argue there never *can* be any physical understanding of the self, or mind. They claim that the mind and the body are quite separate and that mind is necessarily immaterial, an ethereal "something" that just comes to us from somewhere outside and resides temporarily within or alongside the body's "shell." But others, usually of a more scientific bent, have been convinced all along that mind, or consciousness—like anything else—must have some physical explanation. They have argued that its source must be located somewhere in the body, though ideas about exactly where have varied considerably through time and given rise to all sorts of models.

The ancient Greek philosopher Epicurus believed that there were "soul atoms" distributed throughout the body and that they accounted for both consciousness and general vitality, though many early Greeks thought the heart or the breast was the source of these things. Others have supposed that consciousness arose from the function of the liver or lay resident in the blood. According to Hindu philosophers, it was concentrated in *shakras* located along the spine—hence our supposed ability to master it through yoga meditation—while in more recent times Descartes proposed that the meeting point of body and soul was the mysterious pineal gland located in the middle of the brain.

Most of us today who look for a physical seat of consciousness assume that its origin must lie somewhere in the functional capacity of the brain itself. Damage to other bodily organs may result in all manner of troubles, but a sharp blow on the head nearly always results in a loss of consciousness, just as drugs that act on the brain can be seen clearly to alter various patterns of consciousness. Some necessary link, therefore, is assumed to exist between physical states in the brain and conscious or mental states, though the exact nature of that link is still one of the great mysteries of both science and philosophy.

In recent years the favored way of dealing with this enigma has been the rise of functionalism and the tendency to compare the brain to a computer, suggesting that mind, or consciousness, can be equated with the processes that go on inside the computing box. We are what we can do, and what we can do is defined by our circuitry. The computer model still dominates most brain research, and in turn has colored our whole way of perceiving ourselves. We often speak of "needing some input" or of giving "valuable output"; our brains are the "hardware" and our minds the "software"; we get "turned on" and "switched off"; we "blow our fuses" and are "programmed for success (or failure)." The whole of modern biology now operates according to "behaviorial programs" where once there was a sense of purpose, or at least of direction. We think of ourselves as "mind machines."

Certainly the brain is the chief controlling organism in the nervous system, and as such its physical functions include communication, coordination, computation, learning, and memory, all of which it shares to some extent with the operating capacities of our better computers. At that level, analogies between brain functions and computer functions are compelling.

There is an undoubted similarity between the way the brain's complex bundles of neurons are organized and the spaghetti of wiring that makes up the electrical circuitry of a computer, particularly now that computers using parallel processing have been invented. Like the "nerve cells" of a computer, the brain's 10^{10} (ten billion) or 10^{11} (one hundred billion) neurons are also a form of electrical wiring, with various messages passing into and out of the brain by way of electrochemical pulses traveling across the nerve junctions, the synapses. At any one time the brain is literally seething with millions of highly charged neural events, a large proportion of which no doubt underlie our impressive data-processing and computing abilities. But is that what we mean by consciousness? Is computation—for all its diversity and complexity—really all there is to mind? If so, one is tempted to wonder why computers don't have minds.

Certainly they can do very sophisticated things. They can analyze genetic tissue, do complex mathematics, or play chess at a reasonable though somewhat pedestrian standard—but so far no one would argue that any electronic computing device of the sort we can imagine is even remotely conscious. Computers just don't *feel* conscious to us. They lack spontaneity and creativity; they lack imagination; they don't laugh at jokes, enjoy music, suffer pain, or do any of the other things of the sort that we normally associate with the conscious life of the human mind. As one Oxford philosopher put it, "We just do not know what to make of the suggestion that an IBM 100 might be angry or depressed or undergoing an adolescent crisis."[1]

It is conceivable that we might invent sophisticated programs that give computers the *appearance* of such conscious behavior—as in the somewhat spooky case of ELIZA, or DOCTOR, the program designed to simulate Rogerian psychiatric interviews. But as ELIZA's author cautioned, there is a world of difference between a programmed technique or simulation and genuine spontaneity and empathy.[2] It is a form of insanity to think otherwise, though all too often in our mechanized culture it is an insanity that passes for normal.

If we accept the functionalists' equation of being with doing, there is no clear way to argue that something that *behaves* consciously is not conscious. Our whole way of looking at consciousness has become so constricted by the machine model imposed on it that we lose sight of facts linking brain development to consciousness, and seem blind to actual features of our conscious awareness. We become numb to our own experience, and in the process distort it. The danger is that if we continue perceiving ourselves as machines, we may become machines—that is, we may reduce the whole wealth of our conscious life to the far narrower spectrum of thought and behavior that can be written into programs. It is a danger recognized and written about by others,[3] but if we are to surmount it we must find some quite different way of thinking about the mind-brain link, and through that a more human way of perceiving ourselves. In the end, this can only be done through a better understanding of both the physiology of the brain and the physical basis of consciousness.

In fact, the human brain is a complex matrix of superimposed and interwoven systems corresponding to the various stages of evolution, and the self that arises from it is something like a city built across the ages. Its archaeology includes a prehistoric layer, a medieval layer, a Renaissance or Elizabethan layer, a Victorian layer, and some modern buildings. It certainly isn't just a "new town" or "frontier city" built all at once in the last twenty years, as the computer model suggests. Each of us carries within his own nervous system the whole history of biological life on the planet, at least that belonging to the animal kingdom.

In the layer belonging to prehistory we find the one-celled animals such as the amoeba and paramecium, which have no separate nervous system. All their sensory coordination and motor reflexes exist within one cell; our own white blood cells, as they scavenge up rubbish and eat bacteria, behave in the bloodstream much like amoebas in ponds. Simple many-celled animals like jellyfish still have no *central* nervous system, but they do have a network of nerve fibers that allows communication between cells so that the animal can react in a coordinated way; in our bodies, the nerve cells in the gut form a network that coordinates peristalsis, the muscular contractions that push food along.

As the epochs pass, layer upon layer is added to the evolving "city." From the insects onwards we begin to find masses of nerve tissue that carry out more extensive computation, and these increasingly get organized towards the head end. Our withdrawal reflex, which causes us to pull a hand away from a hot stove, involves only the spinal cord and is similar both anatomically and in behavior to that found in earthworms.

With the advent of mammals a forebrain develops—first the primitive forebrain of the lower mammals, ruled primarily by instinct and emotion, and then the cerebral hemispheres with all their sophisticated computing ability, the "little gray cells" that most of us identify with the human mind. Yet drunkenness, the use of drugs like barbiturates and other tranquilizers, or damage to the higher forebrain results in regression to more primitive, more spontaneous, less calculating types of behavior found in lower mammals. Almost the whole of human psychiatry, the actual medical side of treating problems affecting consciousness, concerns itself with regulating the primitive forebrain.

Thus despite the increasing centralization and complexity as the nervous system evolves, the more primitive nerve nets remain, both within the expanding brain and throughout the body. The more recent phases of our evolution have supplanted earlier phases, but not entirely replaced them. The experiences of the amoeba and the jellyfish, of the earthworm and the ant, are all embedded within our own nervous tissue, and with each one of these creatures we share the capacity for consciousness. As Whitehead noted, "The human mind is thus conscious of its bodily inheritance."[4]

Thus whatever consciousness may be, it can't be identical with the higher brain functions permitted by neuron connections in the cerebral cortex. Clearly the *form* our consciousness takes, the contents of our perceptions and our thoughts, is influenced by these connections, but the capacity for consciousness itself, unstructured raw consciousness, must be more basic.

Some animals that are conscious have no cortex at all, others only a very primitive one. Some humans who have vast areas of the cerebral cortex damaged or surgically removed may lose a specific capacity, such as speech or sight or movement, or even memory, but they remain conscious, just as newborn babies are conscious. Consciousness itself, which includes the general capacity for awareness and purposive response, must issue from some physical mechanism that is far more primitive than the developed human brain, from a mechanism that is available to the lowliest amoeba. Understanding how this can be so—finding a basis for consciousness that explains the consciousness of all living (and possibly nonliving) creatures—is crucial for understanding both the place and the *raison d'être* of human consciousness within the general scheme of things.

From a generic point of view these are the questions we must consider when arguing against the computer model of the brain, but there are also phenomenological arguments against it. If we consider carefully certain basic features of consciousness—at least as these are experienced in human beings—it becomes clear that a capacity possessing these features could not *in principle* follow from such a model.

All computer models of the brain share an underlying assumption that the brain itself functions according to the same laws and principles as a vast computing machine—that is, that its separate parts (its neurons) cooperate in an ordered, mechanistic way, following all the determinist laws of classical physics. In such a model, one brain state follows necessarily from another. All we have is one group of static, predictable neurons "looking at" and reacting to other groups. Nowhere in the brain do all these separate groups get integrated. There is no "central committee" of neurons overseeing the whole process, giving it unity and making free, spontaneous decisions. Where, then, in all those trillions of deterministic neural connections and events, is the person we experience ourselves to be? What accounts for the "I" who experiences hunger, decides to eat an apple, and feels pleasure after doing so? How do we even have "an experience" of eating an apple rather than so many scattered impressions of a million different sensory inputs?

The problem has been illustrated by recent work on human vision.[5] When we see an apple, we are immediately aware that it is "an apple," a small, round, red object sitting upright in a bowl on the table some three feet away. There are other associations with the apple in our full conscious perception—that it will satisfy our hunger, that one a day will keep the doctor away, that Eve's decision to eat one was the undoing of humanity. But these are not part of the *visual* perception. That consists of information about the apple's size, shape, orientation, color, and location—each of which, it turns out, is recorded separately by the brain.

The brain does not see "an apple," but rather redness, roundness, smallness, and so on. Information about each characteristic is filed in a different place, on a so-called separate features map and then subsequently on a master map of locations (Figure 5.1). Once the master map has been composed, focused attention takes over, looks at the master map, and sees an apple.

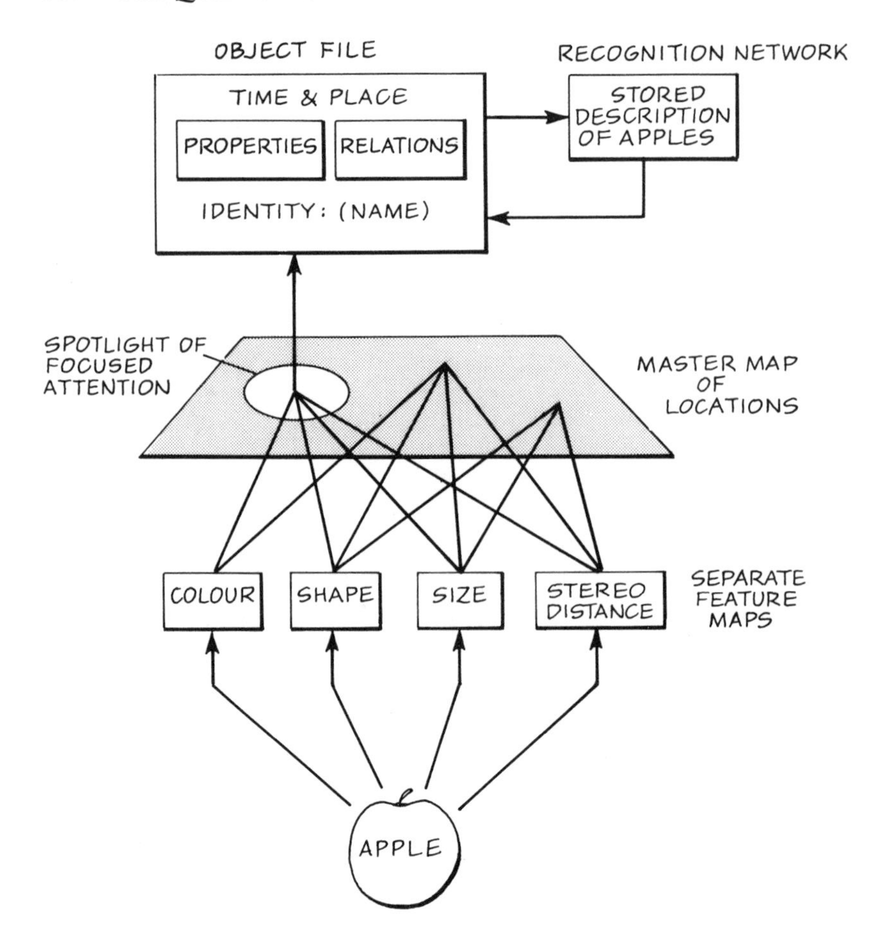

Figure 5.1. A Computer Model of Vision. Information about separate features is collected on features maps and stored on a master map. But focused attention *is required to integrate information on the master map before the recognition network can process the information. What accounts for the focused attention itself?*
Adapted from "Features and Objects in Visual Processing," by Anne Treisman, vol. 255, copyright © 1986 by Scientific American, Inc. All rights reserved.

> Attention makes use of this master map, simultaneously selecting, by means of links to the separate feature maps, all the features that currently are present in a selected location. . . . The integrated information about the properties and structural relations in each object file is compared with stored descriptions in a "recognition network." The network specifies the critical attributes of cats, trees, bacon and eggs, one's grandmothers and all other familiar perceptual objects.[6]

But what *is* this focused attention that integrates information from the master map of perception?

The unity of our conscious experience, the thread of focused attention that draws together the myriad sensory impressions, underlies all other features of that experience. Like the notes of a melody or the many separate features of apples or more general visual scenes, the contents of our consciousness hang together. They form a whole, a "picture." Each part of that whole derives its meaning from the whole and in its own being reflects both the whole and all its other constituent parts. The F sharp I hear is "an F sharp in a Mozart adagio"; it doesn't stand isolated in my awareness. The ash tree I see from my study window is an ash tree at the bottom of my garden on the bank of the Oxford Canal. Its leaves brush the horizon and beckon towards Wytham Woods across Port Meadow. All these things are present to me all at once as I look out of the window. They are, in their wholeness, "my view from the window."

Without this wholeness, this unity, there could be no experience such as we know it, no apples, no gardens, no sense of self (personal identity or subjectivity), and hence no personal will or purposive decision (intention)—all of which are familiar features of our mental life. This unity is the most essential feature of consciousness, so basic to whatever it is that we mean by consciousness that most of us just take it for granted. And yet it is in trying to understand this essential feature of consciousness that we realize just how deeply mysterious consciousness is and why its physics has so far eluded us. There is no comparable unity in any system described by the physics with which we are familiar in our daily lives. The whole corpus of classical physics, and the technology that rests on it (including computer technology), is about the separateness of things, about constituent parts and how they influence each other across their separateness.

If there were no other good reasons to reject the computer model of the brain, the argument pointing towards the unity of consciousness would in itself be damning. As Descartes said when he struggled with the problem of how to explain consciousness in any physical terms, "There is a great difference between the mind and the body, inasmuch as the body is by its very nature always divisible, while the mind is utterly indivisible."[7] This apparently irreconcilable division was one of the arguments that led Descartes to his dualism.

More modern philosophers, while still hoping to find *some* way of explaining consciousness in physical terms, have made much the same point about all classical physical models of the brain, including the computer model.[8] And if the physics of the computer cannot, in principle, give us the physics of consciousness, then it can't be a wholly adequate model of how the brain works, nor, in turn, a very accurate reflection of ourselves and how we function as human beings.

Motivated by the inadequacies of the computer model, some people have proposed a quite different model for thinking about consciousness and the brain, one that is intended to pick up on the theme of unity and account for it in physical terms, and one that makes very different suggestions about consciousness and the self. This is the holographic model, or the "holographic paradigm" as it is sometimes rather grandly described.

A hologram itself is just a special sort of photographic slide that records an interference pattern of light coming from two sources after an initial beam has been split (Figures 5.2 and 5.3). Because the technique for creating holograms is lensless and relies on recording light phase as well as intensity, the resulting slide has a unique way of storing information about any object photographed. The information gathered from any one part of the object is spread all over the slide, so that if some parts of it get destroyed, an image of the whole object can still be projected. However, the greater the area of the slide that gets destroyed, the fuzzier the projected image will be. As Ken Wilber puts it:

> In other words, each individual part of the picture contains the whole picture in condensed form. The part is in the whole and the whole is in each part—a type of unity-in-diversity and diversity-in-unity. The key point is simply that the *part* has access to the whole.[9]

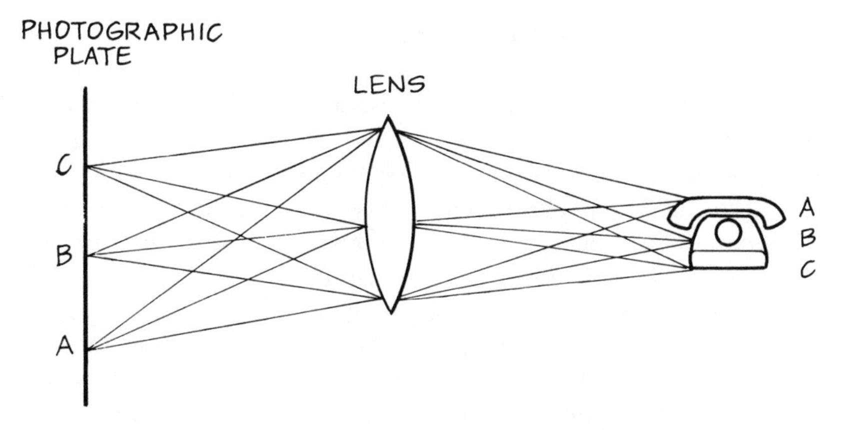

Figure 5.2. Ordinary photographic plate: Light intensity is filtered through a lens, creating an image where partial information is imprinted at separate points on the plate.

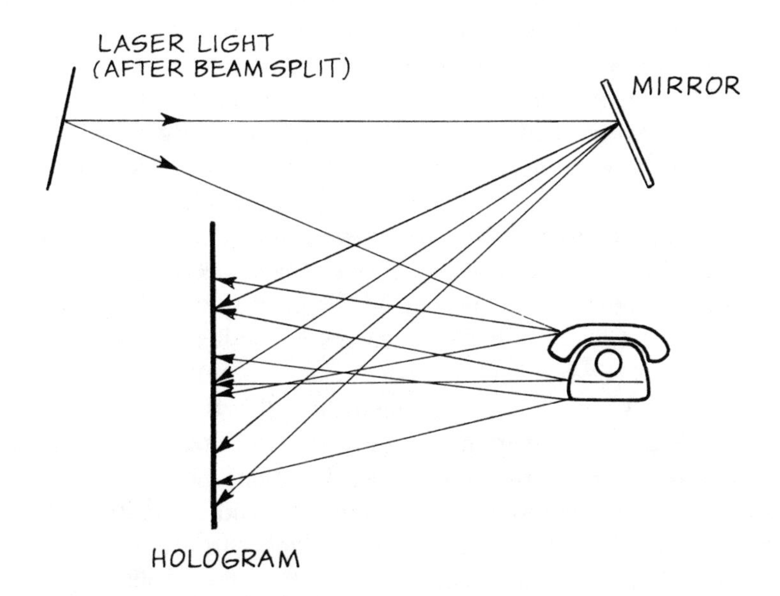

Figure 5.3. Hologram: The interference pattern from split light sources records both phase and intensity. The resulting information from each point of the whole object is stored all over the photographic plate. There is no partial buildup of the image.

There is, as advocates of the holographic model argue, an "eerie similarity"[10] between the ways the brain and the hologram distribute information across the whole system. Each part is privy to information about the whole—though, as critics have pointed out, this property alone doesn't entirely distinguish holography from computation as a model for the brain. The associative nerve nets in the cerebral cortex apparently also function as global information distributors. In fact, their rather messy, higgledy-piggledy wiring design, where everything seems randomly connected to everything else, is the basis for the new computers using parallel processing. This method of computation is very different from that carried out by the older point-to-point computers, which mimic the brain's one-to-one neuron tracts, but it is still computation. This criticism alone, however, would be unlikely to dissuade those who so eagerly embrace the holographic model.

The real impetus behind the holographic model can be traced to a reaction against the computer model and almost everything computers seem to represent, against the whole mechanistic world view and the many forms of alienation and fragmentation that are associated with it. The emphasis of this argument—that the apparently holistic properties of holographic photography mirror the similarly holistic properties of our conscious experience—and the passion that inspires this position hinge on the word *holism* and the extent to which "holistic truths" have been overlooked in mainstream Western culture.

Ever since Plato, the West has stressed the rational and the analytic, the rules by which we form thoughts and make decisions, the "component parts" of our conscious life. The logic of this has led naturally to the computational, or computer, model of the brain. The cost of this model, however, has been the overlooking of another side to human knowing and experience, what might be called the intuitive side, the side that draws on wisdom, imagination, and creativity. In modern neurophysiological terms, these two sides of our mental life have been spoken of as the right-brain/left-brain split, and our culture as a left-brain culture.* Using an equally good metaphor from quantum physics, we might speak of this situation as a particle/wave split and say that our culture has emphasized the particle aspect of the mind.

---

*The capacity for analysis and logical thought issues almost exclusively from the functional capacity of the brain's left hemisphere.

The "holists" want to emphasize the wave aspect of experience, the extent to which every element of consciousness—indeed every element of reality itself—is related to everything else. The whole is something greater than the sum of its parts, or as David Bohm, one of the chief proponents of the holographic model, puts it, reality is an "undivided wholeness."[11] Everything—and everyone—is so integrally related to the rest that all talk of individuals or separation is a distortion of truth, an illusion.

This latter-day holism has its antecedents in both East and West. As the Buddhist Diamond Sutra expresses it:

> In the house of Indra there is said to be a network of pearls so arranged that if you look at one you see all the others reflected in it. In the same way, each object in the world is not merely itself but involves every other object, and in fact *is* every other object.[12]

Much the same metaphor occurs in our Western tradition, as in the Great Chain of Being, which links the microcosm and the macrocosm by asserting that each small bit of reality contains the whole, or in Spinoza's philosophy, which emphasizes that everything in the world is made of one substance. Those who suggest the hologram as a model for the brain are attempting to put such metaphors on a scientific footing.

Both the holographic paradigm in general and the holographic model of the brain in particular have their attractive qualities. As a metaphor accessible to the modern mind, the hologram plays a useful role in stressing the aspects of consciousness and reality that stem from relationship and process, and thus serves to remind us that we are all parts of some larger whole. But even as a metaphor it goes too far in some respects, being as extreme in its own emphasis on the wavelike side of being as mechanism and the computer model are in their emphasis on the particle side. Reality as we know it consists of both waves (relationship) and particles (individuality), just as the experience we recognize as human mental life consists of both immediate consciousness (unity and integration) and computation (thought, structure). A really adequate model of the nature of consciousness and its relationship to the brain must be able to account for both.

As an attempt to put the unity of consciousness on a sound scientific footing, the holographic model fails on at least two counts. In the first place, like the computer model, it can't account for the "I" of consciousness. If "the brain is a hologram perceiving and participating in a holographic universe,"[13] who is looking at the hologram? The hologram itself is nothing but an unusual sort of photograph, which can't on its own be doing any perceiving. So in asking who or what supplies the consciousness (the "spotlight of focused attention"), one is driven to assume either that this must come from outside, as the dualists have argued all along, or that the physics of the hologram can account for the unity of conscious perception. Obviously, it can't.

Since holograms are constructed by recording the interference patterns of light waves, and such interference patterns are an entirely classical effect described by standard electromagnetic theory, holograms themselves are classical systems. That is, despite their ability to store information about a whole object in each part of the photographic plate, they are ultimately divisible into parts. They *are* so many separate marks on a plate, which, when enough of them are seen together, represent a whole. But one can always reduce the number of marks perceived to a point where they convey nothing of the whole.* This is not the kind of holism required to explain the unity of consciousness, and in that crucial respect the hologram is no better than any other classical model for explaining the physical basis of consciousness. In fact, there is very little to distinguish it from the computer model of visual perception discussed earlier (see Figure 5.1, page 68). That model's master map of locations might well *be* a hologram constructed through the parallel processing of visual data. Just as "the spotlight of focused attention" was a crucial missing link in the visual process, so it is in the holographic model of more general brain function.

Nonetheless, as increasing numbers of people feel a pressing need to find some way beyond the lonely isolation and general alienation imposed by the strong mechanistic strain in our culture, the desire for some sort of holism is in the air. As one philosopher has characterized it lightheartedly, holism is all "organic and fuzzy and warm and cuddly and mysterious."[14] Hence the popularity of the holographic paradigm, of David Bohm's picture of undivided wholeness, and the general revival of Eastern mysticism with its emphasis on the oneness of all

*"Each component of the electric field at each point on a space-like surface is a separate degree of freedom, and all these infinitely-many degrees of freedom can in principle be assigned independent values." (Abner Shimony, 1981, p. 435.)

things. All attempt to transcend our separation from each other and from the world at large.

In many ways this book can be seen as part of that general holistic movement, though I see no need to rest it on the insights of Eastern mysticism, and throughout I shall argue that "undivided wholeness" is only one side of a comprehensive picture of reality and the place of consciousness within it.

But if holism is to have any real meaning, any teeth, it must be grounded in the actual physics of consciousness, in a physics that can underpin the unity of consciousness and relate it both to brain structure and to the common features of our everyday awareness. I think that to achieve that, we must turn to quantum mechanics.

CHAPTER 6

A QUANTUM MECHANICAL MODEL OF CONSCIOUSNESS[1]

> We may well now ask whether the close analogy between quantum processes and our inner experiences and thought processes is mere coincidence. . . . The remarkable point-by-point analogy between thought processes and quantum processes would suggest that a hypothesis relating these two may well turn out to be fruitful. If such an hypothesis could ever be verified, it would explain in a natural way a great many features of our thinking.
>
> —David Bohm
> *Quantum Theory*

In earlier chapters we saw that in some very important ways consciousness is an issue for quantum mechanics. But it is also true, certainly at the level of analogy, that quantum mechanics is no stranger to consciousness. It was nearly forty years ago that David Bohm first drew out the many striking similarities between the behavior of our thought processes and that of some quantum processes.

Who, for instance, has not had the experience of entertaining a vague train of thought, only to find that the act of concentrating in order to bring it into better focus somehow changes the original sequence or "flavor"? Like the electrons governed by Heisenberg's Uncertainty Principle, which are never the same again once they've been looked at (measured), a thought that has been highlighted through attention is different from the vague musing that preceded it. We might say the focused thought has "position," like the particle aspect of an electron's two-sided nature, whereas the vague musing had "momentum," like the wave aspect. We can never experience (measure) both simultaneously.

Then, too, just as quantum systems are essentially unified, so are our thought processes. I can no more separate the peculiar charm of my daughter's toothless smile from the fact that she is my daughter than a physicist can separate the electron he is measuring from the instrument with which he is measuring it. The meaning of each—in the case of the electron, its mode of being—depends on its place within a relationship, on its context. Thus, as Bohm says:

> Thought processes and quantum systems are analogous in that they cannot be analyzed too much in terms of distinct elements, because the "intrinsic" nature of each element is not a property existing separately from and independently of other elements but is, instead, a property that arises partially from its relation with other elements.[2]

And finally, there is an intriguing parallel between the way logic helps to structure and focus our otherwise "indeterminate," flowing thought processes and the way the laws of classical physics make it possible to describe the everyday world of separate objects and causal relationships that overlie and are the limit of quantum-level processes. Without this classical limit there would be no solid, "real" world; without logic there would be no way to express our thoughts clearly, no way to test them against the outside world.

"Thus," according to Bohm, "just as life as we know it would be impossible if quantum theory did not have its present classical limit, thought as we know it would be impossible unless we could express its results in logical terms."[3]

That there is a vital link between thought processes and quantum processes, between ourselves and electrons, is the underlying assumption of this book, and the many analogies between the two are both tantalizing and suggestive. Analogy has been a powerful tool in the development of both philosophy and scientific thought, and on the strength of that alone there is a powerful case for drawing out the relationship between quantum processes and much of everyday life.

But if it really were possible, as Bohm himself suggested in those early days, to go beyond analogy, to get beyond saying that thought processes are *like* quantum processes and go further on to *explain* consciousness in terms of quantum mechanical features in the actual structure and functioning of the brain, we would have taken a truly revolutionary step. Not only would we be very much closer to understanding the physical basis for many aspects of both individual and group psychology, but we would also have gone a long way towards understanding our relation to Nature and the material world.

The construction of a model that demonstrates one possible way in which consciousness could be seen to function according to the laws of quantum mechanics is the purpose of this chapter. Having laid a tenable foundation here, we will be able in later discussion to draw out some of the philosophical and psychological consequences of such a close bond between the dynamics of the self and those of the electron.

At the time Bohm first described the analogies between thought processes and quantum events, it would have been impossible to go much further. Neither neurobiology nor quantum physics was sufficiently developed to demonstrate how any aspect of one could very easily be explained in terms of the other. Most crucially, the whole explosion of thought—and bewilderment—that has followed in the wake of proof that nonlocal correlation effects ("action-at-a-distance," in very crude terms) exist between particles apparently separated in space and time was yet to come. Without that, and the even stronger unifying effects found in some large, ordered structures like lasers and superconductors, a physical understanding of consciousness is impossible; with them, a quantum mechanical approach becomes attractive.

As the inadequacies of the computer model and the holographic model illustrate, the central problem of understanding consciousness in physical terms, the rock against which all previous theories have broken, is the problem of the unity of consciousness, the distinctive indivisibility of our thoughts, perceptions, and feelings. Without it there could be no experience such as we know it and no self having that experience. No process in classical physics gives rise to that sort of unity, and until fairly recently it was not a major theme in quantum physics. But now that special sorts of specifically quantum mechanical unity are recognized, both physicists and philosophers have begun to wonder whether they might not have some meaningful relevance to the unity of consciousness. Oxford's Roger Penrose puts the case for them all:

> Quantum physics involves many highly intriguing and mysterious kinds of behaviour. Not the least of these are the [nonlocal] quantum correlations which can occur over widely separated distances. It seems to me to be a definite possibility that such things could be playing a role

> in conscious thought modes. Perhaps it is not too fanciful to suggest that quantum correlations could be playing an operative role over large regions of the brain. Might there be any relation between a "state of awareness" and a highly coherent quantum state in the brain? Is the "oneness" or "globality" that seems to be a feature of consciousness connected with this? It is somewhat tempting to believe so.[4]

An analogy to the kind of quantum correlations Penrose is suggesting here would be a group of musicians, playing and recording in separate rooms, who nonetheless conspire to produce a harmonious result. Or the phenomenon of the quantum twins we discussed in Chapter 2, who, though separated by ignorance and thousands of miles, nonetheless lead entirely synchronistic lives. Such quantum systems do seem to resemble the way separate neurons all over the brain cooperate to produce a unified state of conscious awareness, though such an observation in itself adds little to Bohm's earlier analogies.

The first substantive evidence that there is at least a channel of communication between the world of quantum physics and our perception of everyday reality was found nearly half a century ago. At that time biophysicists working on the retina discovered that nerve cells in the human brain are sufficiently sensitive to register the absorption of a single photon (mirroring the passage of an individual electron from one energy state within the atom to another)—and thus sensitive enough to be influenced by the whole panoply of odd quantum-level behavior, including indeterminism and nonlocal effects.

Further experiments proved that quantum indeterminancy is built into the functioning of the brain itself, through random variations in the chemical concentrations surrounding nerve junctions (neuron synapses). These concentrations determine the levels at which neurons "fire"—i.e., make electrical contact with other neurons—and even very slight, quantum-level variations affect the firing potentials. Indeed, the levels at which neurons fire vary according to definite statistical law, just like any other quantum process. Of the brain's 10^{10} neurons, some 10^7 are believed sensitive enough to register quantum-level phenomena at any one time. Still, the firing of individual neurons is a very long way from explaining whatever complex processes might be associated with the brain's conscious activities.

The necessity of a quantum mechanical approach to consciousness itself was first worked out in detail in 1960 by Ninian Marshall in a paper on telepathy and memory.[5] Marshall's argument was that the deterministic laws of classical physics left no room for the free play of thought processes, free will, or intention—all of which we take to be common features of consciousness. No physical brain mechanism obeying the determinist laws of classical physics could account for freedom of thought or will, or for any of the free actions that might follow from them.

Much the same sort of point was pursued more recently by the Russian physicist Yuri Orlov. He argued that in any kind of doubt resolution or creative thinking, quantum indeterminism and superimposed probability states (virtual states) must be playing a role in the brain's openness to all the potentialities latent in consciousness—for example, our ability to see many possibilities all at once.

> The described mechanism [quantum indeterminism] . . . gives a key to the understanding of *creative thinking,* when a person states or depicts "what in fact does not exist." According to our approach, the person potentially "sees" several versions simultaneously without completely realizing any of them, and then one version "pops up" (materializes) as the result of a free choice.[6]

The playing out, simultaneously, of many different—and eventually mutually exclusive—possibilities is reminiscent of the quantum hussy we met earlier when discussing virtual states. Just as her free love finally had to give way to commitment, so our free play of thought and imagination must at some point resolve itself into a settled idea. Only one of a given set of quantum possibilities can exist in the "real world," but before that one materializes, what fun the quantum world allows us to have!

But if, as people like Penrose, Marshall, and Orlov suggest, the physical basis of consciousness is some sort of quantum mechanical phenomenon, with all the freedom that implies, a great deal remains left unexplained. What sort of quantum process might it be, for example, and what properties of the brain could possibly sustain it? It is only by attempting to answer such basic questions that a model of consciousness resting on quantum physics could have any real meaning.

If we accept that the unity of consciousness is the single most significant sticking point when it comes to explaining consciousness in terms of known physics, we can see that certain features of that unity might offer hints about the nature of any underlying physical process. The background state of all consciousness, for instance—the blackboard on which various individual thoughts and perceptions are written—is what physicists call a steady state. It is uniform in space and persistent in time, qualities that are necessary for consciousness to do the job that it does. Just as we couldn't write much of a message on a bumpy or short-lived blackboard, so the particular contents of our conscious awareness would not easily be distinguishable if the general background against which they are set were not a steady state. As ethologist John Crook has said, "The orderliness of awareness—its apparent stability within time—is what gives us the feeling that we live in a world rather than within experiences conjured up by capricious senses."[7]

But this very "orderliness of awareness" considerably limits the choice of underlying physical explanations, as can be gathered from the failure of all attempts to explain consciousness in classical terms. Our consciousness has the character of unbroken wholeness. It hangs together and allows our experience to do so. This kind of settled uniformity is rare among dynamic processes in nature, but it does occur in materials that exist in "condensed phases." The physics (and the physiology) of condensed phases thus seems a candidate worth investigating further to see if it can suggest an explanation for how consciousness could arise in brains.

A phase is a "state," or condition of something, of some material system, just as a "teenage phase" or "a bohemian phase" is a possible "state" of the psyche. In natural materials it refers to the amount of order existing in a given system. Water, for example, has three phases—gaseous (steam), liquid (water), and solid (ice)—and each displays a greater order among its molecules than the last. The solid, an ice crystal, is one very familiar example of a loosely structured condensed phase, as are salt crystals and sugar crystals.

There are other, reasonably familiar examples of more properly structured condensed phases in physical Nature—ordinary magnets, superfluids, superconductors, laser light, electric currents in metals, and sound waves in crystals. The property that all these things have in common is some degree of coherence, such that the many atoms or molecules that make up the substance suddenly (or gradually) behave as one.

Imagine, for example, a large number of electromagnetic compasses lying on a table in a shielded room. Because the room is shielded, the compass needles point in no particular direction, and if the table is jiggled, they swing randomly in every direction possible. A physicist wanting to describe the motions of the needles would have to write many equations—one for each needle. But if the electromagnetic energy in each compass is increased, the needles begin to exert a pull on each other, and slowly they line up into a uniform pattern. At the point where the electromagnetic current becomes strong enough to outweigh the effect of jiggling the table (the equivalent of thermal noise in a real system, where heat makes the molecules jiggle about), it would have the dramatic effect of making all the needles point in the same direction (Figure 6.1). All the separate compasses would then behave like one supercompass, and the physicist could write just one equation to describe the motion of the ensemble. We would say that the compass needles have gone into a condensed phase.

But if any of this is relevant to consciousness, one could be forgiven a first reaction of demanding to know how the neurons of the brain could possibly get into a condensed phase. Living cells appear to be different in nearly every respect from something like a magnetized compass needle, and even though it is an argument of this book that the living and the nonliving worlds interact in some previously overlooked way, when the crunch comes it seems strange at first to think of the actual mechanics of a similar process in the brain. What kind of neurobiological mechanism would be required to "line up" neurons (or some constituent of them) in the way the compass needles of the example were lined up by their own internal magnetic fields? And is such a mechanism feasible?

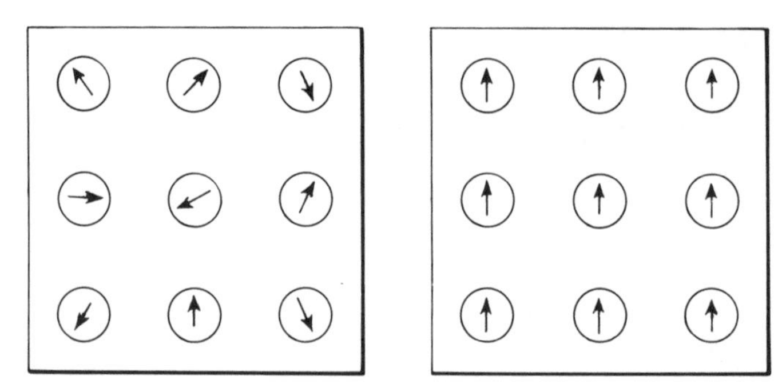

Figure 6.1. Without electromagnetic current, the shielded compass needles point randomly (left); *with the electromagnetic current, they line up* (right).

Various people have suggested that consciousness might depend on the brain's somehow taking on the characteristics of a superfluid or a superconductor.[8] But while either of these would satisfy the requirement of being the kind of highly ordered state found in a condensed phase, it would scarcely satisfy the demands of feasibility. Superfluids and superconductors exist only at very low temperatures, whereas brains, as we all know, function at normal body temperature. If the physics of condensed phases is to prove relevant to consciousness, then there would have to be some such mechanism that functions at normal body temperature. And in fact there is one. The "pumped system" first described by Professor Herbert Fröhlich of Liverpool University in England some twenty years ago, and known to exist in biological tissue, seems to satisfy all the necessary criteria.[9]

Fröhlich's pumped system is simply a system of vibrating, electrically charged molecules (dipoles—positive at one end and negative at the other) into which energy is pumped. As they jiggle, the vibrating dipoles (molecules in the cell walls of living tissue) emit electromagnetic vibrations (photons),* just like so many miniature radio transmitters. Fröhlich demonstrated that beyond a certain threshold, any additional energy pumped into the system causes its molecules to vibrate in unison. They do so increasingly until they pull themselves into the most ordered form of condensed phase possible—a Bose-Einstein condensate.

The crucial distinguishing feature of Bose-Einstein condensates is that the many parts that go to make up an ordered system not only *behave* as a whole, they *become* whole; their identities merge or overlap in such a way that they lose their individuality entirely. A good analogy would be the many voices in a choir, which merge to become "one voice" at certain levels of harmony, or the plucking of the many strings of several violins to become "the sound of the violins." This merging of identities is crucial to giving any physical accounting of the way consciousness draws together various "subunities" within experience.

While it is certainly the case that one person can possess two or more "islands" of consciousness—for instance, the experience of carrying on a conversation while continuing to drive a car—the experience of having one general field of consciousness is almost universal. There is not one someone sitting here being conscious of an ash tree on the bank

*To be exact, *virtual* photons—photons with a limited range of interaction.

of a canal, another someone noticing the rattle of a nearby train, and yet a third someone being aware of mild backache. They are all one person, "me."

Yet, for the person who entertains each of these different "mini-conscious" experiences to *be* the same person, for there to be one integrated self experiencing them all together, something must account for the unity of the separate brain states associated with each of the contributing elements of the experience. In each one of these states at any moment, there are at least one hundred different bits of information. To bring all this together, to achieve the degree of unity required, necessitates that the separate brain states attending to each element become *identical. All* their properties and *all* their information must entirely overlap. This kind of unity is found only in Bose-Einstein condensates.

And it is only in such condensates, where individuality breaks down, that we can find distinctively quantum mechanical effects in large-scale systems. A quantum physicist would say that the wave functions of the previously individual bits have overlapped—they've become indeterminate in their spatial location so that each one spreads itself *all over* the whole, just as the indeterminate quantum hussy lived with all of her lovers at once, or the alive/dead cat belonging to Schrödinger spread out his ambiguous being to fill the entire box that enclosed his secret.

Such large-scale quantum synchronicity accounts for the special properties of lasers, superfluids, and superconductors, but the importance of the type found in Fröhlich systems is that it exists at normal body temperature. Indeed, it is found *only* in biological tissue, where the vibrating charged dipoles within cell walls emit signals of microwave frequency as they vibrate. Such frequencies commonly exist in and have an effect on biological tissue[10] (the growth rates of yeast cells, for example, are influenced by microwave radiation). So far, however, the *reason* why living cells should generate and respond to microwave radiation, and thus have the capacity to contain Bose-Einstein condensed phases within their walls, remains a mystery—though one calling for some explanation. As Fröhlich himself says, "Biological systems have . . . developed to fulfil a certain purpose, and it is permissible, therefore, to ask for the purpose of a certain excitation."[11]

One physicist suggests that the purpose of the microwave-producing vibrations in living cells might be related to the way living systems, as opposed to nonliving ones, create order out of Nature's chaos and confusion.[12] When cell membranes vibrate sufficiently to pull themselves into a Bose-Einstein condensate, they are creating the most coherent form of order possible in Nature, the order of unbroken wholeness. This might be the mechanism by which life violates the Second Law of Thermodynamics (entropy), according to which all inanimate systems are destined to degenerate into chaos.

Other biophysicists—some working as colleagues of Professor Fröhlich, others independently—have found evidence for the same sort of biological coherence, though their research suggests that it is a coherent ordering of photons in the range of visible light, rather than (or in addition to) those in the microwave range.

German biophysicist Fritz Popp has discovered that living cells emit a weak "glow," which is evidence of photon radiation, and suggests that the presence of what he calls coherent biophotons may play some crucial role in cell regulation.[13] Scientists working independently in Japan have discovered the same effects, which they believe "are clearly associated with a variety of vital activities and biological processes."[14] At least one Polish biophysicist has found evidence of the coherent ordering of photons in DNA itself,[15] as have Popp and one of his German colleagues.[16]

Evidence for coherent states (Bose-Einstein condensates) in biological tissue is now abundant, and the interpretation of its meaning is at the cutting edge of exciting breakthroughs in our understanding of what distinguishes life from nonlife. *I think that the same Bose-Einstein condensation among neuron constituents is what distinguishes the conscious from the nonconscious. I think it is the physical basis of consciousness.* *

If, however, we want to suggest that Bose-Einstein condensation is the physical basis of consciousness, we must look for the necessary features of a Fröhlich-style system in the brain. I suggest that the electrical firing that constantly takes place across neuron boundaries whenever the brain is stimulated might be providing the energy required to jiggle molecules in the neuron cell walls, causing them to emit photons. By way of such signals the molecules in any given cell walls, and in thousands of nearby ones, could communicate with each other

*Earlier authors have suggested that Bose-Einstein condensation in the brain might be the physical basis of memory, though they could not find a suitable mechanism. See C.I.J.M. Stuart et al., 1979.

in a "dance" that begins to synchronize their jiggling (or their photon emissions). At a critical frequency they would all jiggle as one, going into a Bose-Einstein condensed phase. The many dancers would become one dancer, possessing one identity.

At that crucial point, the point of a phase shift into the condensed phase, the movements of the synchronized molecules within neuron cell walls (or photons emitted by them) would take on quantum mechanical properties—uniformity, frictionlessness (and hence persistence in time), and unbroken wholeness. In this manner they would generate a unified field of the sort required to produce the ground state of consciousness. The phase shift, then, is the moment when "an experience" is born.

Just one of the many interesting implications of thinking about consciousness in terms of a Fröhlich system is that it lends support to the view that some rudimentary consciousness may well be a property of *all* living systems. If a Fröhlich-style Bose-Einstein condensate can be found in yeast cells, then it would most likely follow that any biological tissue comprising at least one cell—plant or animal—would have the fundamental unifying capacity required to support some sort of conscious awareness. A smaller Bose-Einstein condensate, however, wouldn't have as many possible states (excitations). Its range would be limited—thus a snail would have a much more limited consciousness than we do.

Indeed, there is no reason in principle to deny that *any* structure, biological or otherwise, that contained a Bose-Einstein condensate might possess the capacity for consciousness, though the *sort* of consciousness it would have and what could be achieved with it would depend on the system's overall structure. This leaves open the possibility of conscious computers, and of course raises the question of alien consciousness in general.

In familiar, terrestrial higher animals like ourselves, the electric fields across neuron cell walls are constantly varying as a result of fluctuations in the amount of energy being pumped into the system. These fluctuations are due to chemical changes in the blood, such as raised or lowered blood sugar, and to outside stimulation. It follows from this that the strength of consciousness would also vary, with more or fewer (protein or fat) molecules being pulled into or out of the condensed phase. This accords with our everyday experience, where we are more conscious at some times than at others (for instance, in a state of high concentration versus a state of sleep). It also accords with what we know about the effect, or lack of it, of brain injuries on consciousness.

If, as the computer model of the brain suggests, consciousness arose from the brain's computation mechanisms—the many billions of individual neurons connected with each other like telephone cables—it should then, like a telephone system, suffer interrupted function when one or more of the cables gets broken. This does indeed happen to certain specific functions of the brain after injury—damage to the optic area destroys sight; to the auditory area, hearing; and so on. But consciousness itself doesn't suffer in the same manner from such localized injuries.

It is only after a very massive brain injury destroys large sections of the brain (or under the influence of drugs, such as anesthetics) that consciousness is sufficiently affected to lose its holistic property, as we would expect if consciousness is a nonlocal quantum phenomenon. In a theory based on something like Fröhlich's pumped system, the most fundamental aspect of consciousness—its capacity for unified awareness—has nothing whatever to do with individual neuron connections in the brain.

In the quantum mechanical model of consciousness being suggested here, the vibrating molecules in the neuron cell walls (or photons associated with them) that give rise to a Bose-Einstein condensate account only for the ground state of our awareness, the "blackboard" on which things (perceptions, experience, thoughts, feelings) are written. The "writing" itself would be supplied from a wide range of sources—the genetic code, memory, synaptic activity in the brain, and all those phylogenetic echoes resonating within the nervous system. Each of these would appear individually or in some combination as excitations of the underlying condensate, as patterns within it, like waves on the sea or bubbles on the surface of a pot of boiling stew.[17] And it would be these patterns, the mathematics of which are actually the mathematics of a hologram,[18] that we recognize as the familiar contents of consciousness (Figure 6.2). Interestingly, Descartes believed that perceptions were excitations of the underlying soul.

This model, along with the suggestion that excitations of the Bose-Einstein condensate account for the recognizable patterns of our conscious life, also suggests an interpretation for the as yet mysterious EEG patterns recorded when electrodes are attached to the skull to measure

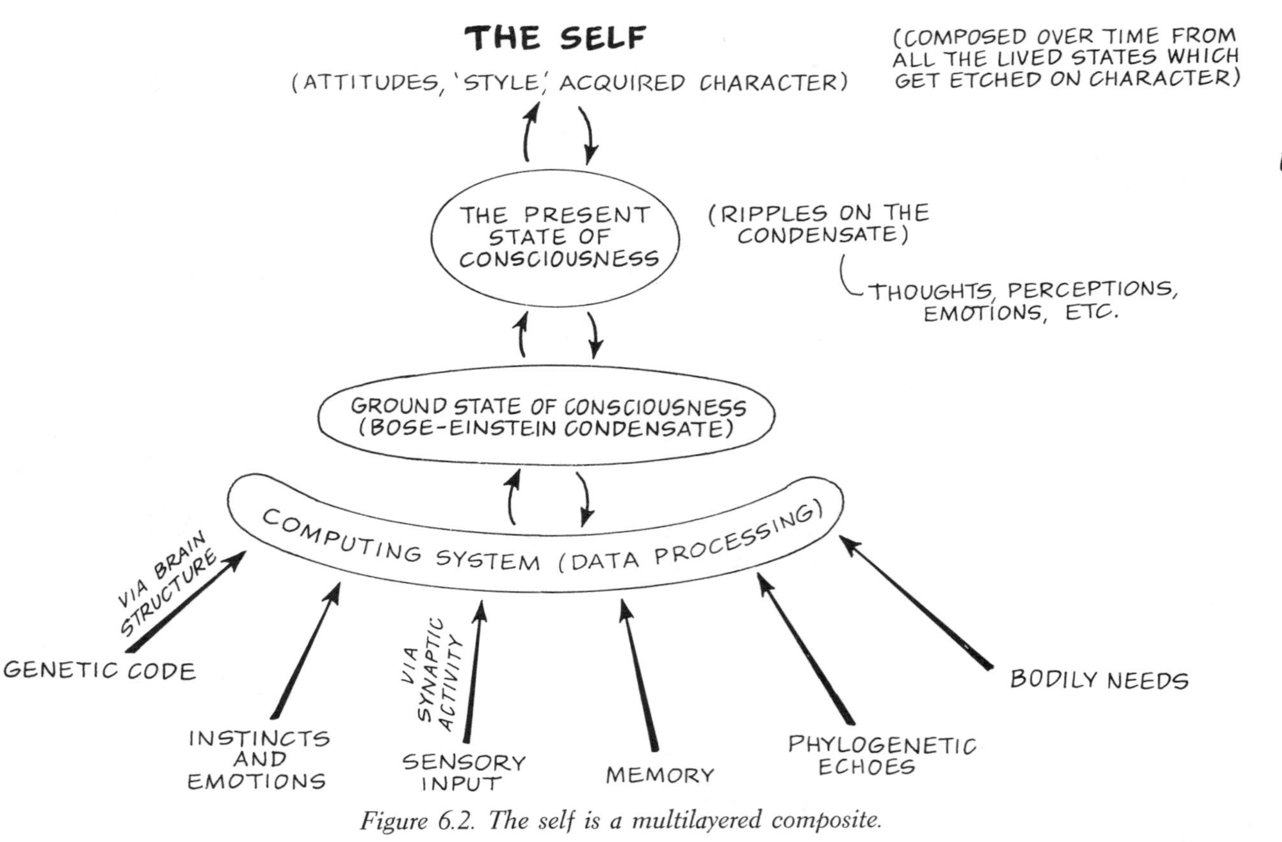

Figure 6.2. The self is a multilayered composite.

brain activity. The typical wave patterns seen on the EEG, which are believed to record subthreshold (prefiring) oscillations in neuron cell walls,[19] vary depending on one's state of consciousness and the activity in which the brain is engaged. Four distinct patterns have been recognized: alpha, beta, delta, and theta (Figure 6.3).

In the normal adult human brain, the beta waves, which are associated with organized, conceptual thinking, dominate the EEG pattern during waking hours; delta waves are found when the brain is in a state of deep, dreamless sleep; theta, during dreaming sleep; and alpha, in a state of deep relaxation, when the brain is fully awake but not focusing on any particular idea.

Each overall EEG pattern is stable, although—as is true of waves in general—the individual neurons that embody it change from moment to moment. In both the whole-scalp EEGs and, more dramatically, in those done on two individual neurons involved with the same visual stimulus, the wave patterns representing excitation are synchronized, suggesting that a long-range coherence binds the firing patterns of individual neurons.[20] On the basis of any classical interpretation of neuron connections, it is difficult to explain this, but the suggestion

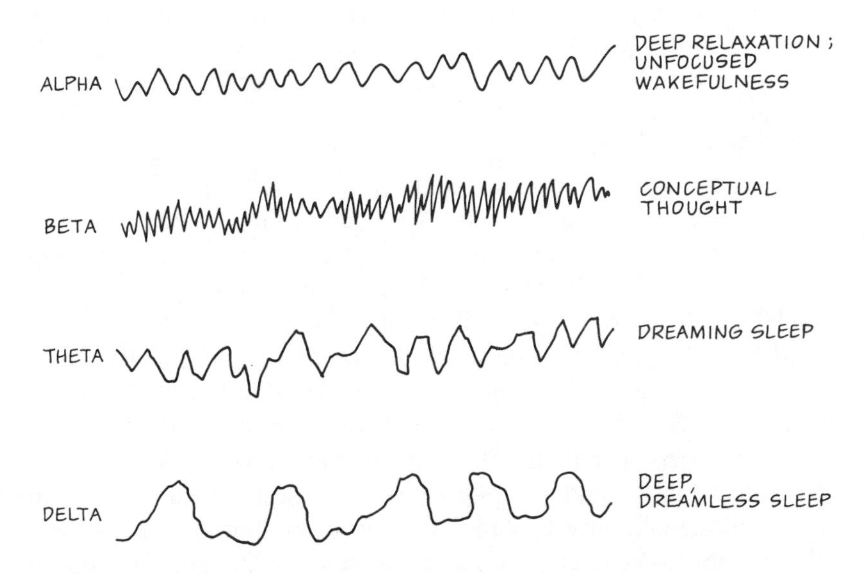

Figure 6.3. EEG patterns of brain activity. Could these be mirroring excitations of the Bose-Einstein condensate?

that the brain has a quantum integrating system makes interpretation easier.

In the model of consciousness I am suggesting, the brain has two interacting systems: the coherent Bose-Einstein condensate associated with consciousness, and the computerlike system of individual neurons. The electrical activity observed on the EEG may be a bridge between the two—if either system is excited, it would produce an electrical field that acts on the other.[21] But because of the quantum factor, the excitations would always be integrated, i.e., coherent.

A quantum mechanical model of consciousness, then, gives rise to a picture of our overall mental life that is neither entirely like a computer nor entirely like a quantum system—indeed, not entirely "mental." What we recognize as our full-blown conscious life, using *conscious* in its common vernacular sense, is actually a complex, multilayered dialogue between the quantum aspect (the ground state) and a whole symphony of interactions that cause patterns to develop in the ground state. These patterns encompass interactions with our computing facilities in the cerebral cortex, with our instinctual and emotional capacities in the primitive forebrain, with our appetites and our twitches (or pains), with a whole host of activities going on in the body, and to some extent with the conscious lives of other people and creatures. It is the quality of playing by the various members of this orchestra that ultimately determines the overall quality and content of the music played—our conscious life.

Whether or not any existing theory, such as that applying Fröhlich's pumped systems or Popp's coherent photons to the problem, proves correct, the very existence of a viable quantum mechanical model of consciousness is already pregnant with far-reaching philosophical implications. The unbroken wholeness that is a prerequisite for any such model, and hence the loss of individuality of its constituent parts, bears on the whole question of personal identity and group relations.

Then, too, any quantum mechanical model is necessarily a physical model. We can, therefore, assume that the phenomena of consciousness (awareness, perception, thought, memory), along with those of physics, chemistry, and biology, belong to the order of Nature and can be experimentally investigated. This way of looking at consciousness also implies that consciousness and matter are so integrally bound up with each other that either consciousness is a property of matter (as in panpsychism) or else, as Nagel suggests (Chapter 4), consciousness and matter arise together from the same common source—in our terms, from the world of quantum phenomena.

Either view takes consciousness out of the realm of the supernatural and makes it a proper matter for scientific inquiry. It challenges the widely held dualist assumption that consciousness and matter ("mind" or "soul," and body) are entirely separate phenomena, each evolving in its own way and only accidentally touching the other in this imperfect world of ours. If it were to be proven that consciousness is indeed a quantum process, the long-standing claims of dualism would be challenged in a way more profound than ever before. We are now in a position to reassess the whole question of how the mind and the body relate.

CHAPTER 7

MIND AND BODY

> I rightly conclude that my essence consists in this alone, that I am a thinking thing. . . . And although perhaps . . . I have a body with which I am closely conjoined, I have, on the one hand, a clear and distinct idea of myself as a thinking, non-extended thing, and, on the other hand, a distinct idea of my body as an extended, non-thinking thing; it is therefore certain that I am truly distinct from my body, and can exist without it.
>
> —DESCARTES
> *Meditations*

When my young daughter asked me what her "soul" was, I heard myself telling her that it was the most essential part of herself, the part that made her truly "her," and that it was different from her body. Had she been sophisticated enough to ask about her "mind," I feel certain I would have responded in kind—in spite of myself. Despite all my thinking about the subject and my rational convictions to the contrary, I am a good Cartesian at heart, and when I struggle to give my children explanations about something so basic as the mind/body or soul/body relationship in terms they can understand, I find myself drawing on some deep pattern of belief laid down in my own childhood and strengthened by the whole tenor of my education. I suspect it is the same for most people, even those who have never read, or even heard of, Descartes.

Most of us *feel* that our minds (or souls) and our bodies are somehow essentially different, whatever we might otherwise think in our more rational reflections. Our experience is of a self that *has,* or a self *within,* a body. We feel the self to be deeply private, tucked away, an intangible something that peeks out at the wider world beyond and that might enjoy all manner of capacities and freedoms but for the body's limitations. At the best of times we soar beyond this fleshy constraint. We are healthy despite its diseases, young despite its gray hair and wrinkles, "pure" despite its "corruption." At the worst of times we sink to its level and cry out in despair.

"O wretched man that I am!" cried Saint Paul, "who shall deliver me from the body of this death? . . . So then with the mind I myself serve the law of God; but with the flesh the law of sin."[1] Paul's talk of vile bodies and mortifying their deeds left an indelible mark on the whole development of Christianity and hence on the psyche of Western man. But he, too, though so much our teacher, was himself the product of his own education and culture—in his case the sentiments expressed in Plato's *Phaedo* and *Republic* nearly half a millennium before and carried on forever more in the Platonist and Neoplatonist traditions.

"So long as we keep to the body and our soul is contaminated with this imperfection," complained Socrates, we are lost to our pursuit of truth. "The body fills us with loves and desires and fears and all sorts of fancies and a great deal of nonsense, with the result that we literally never get an opportunity to think at all about anything."[2] For his part, Socrates gratefully seized the hemlock and looked forward to being happily dead, after which his immortal soul could get on with what really mattered.

However much our modern reason might wish to shake off the mind/body or soul/body dichotomy, this deep cultural conditioning holds us in its grip, not least because the physics of the last three hundred years supports it. Ever since Descartes brought dualism to its most succinct and forceful expression in the seventeenth century by resting it on the new mechanical concepts of mass and matter, subsequent philosophers have tried in vain to argue for any viable alternative. Ordinary people have had the same problem. Given our everyday, essentially Newtonian notion of what matter is, and hence what bodies must be, there is no clear way to see how they could be anything like minds.

Newtonian physics took the older Platonic and Christian notion that matter was something "base, inert, shapeless and 'plump,' "[3] and sharpened it considerably. Matter was something that had weight and extension; it was essentially atomistic, consisting of tiny corpuscles that behaved like so many billiard balls. Hence it was solid, it influenced other matter mechanically by touching it, and, most important of all as a break with the past, it was wholly mindless.

Matter had no purposes or intentions. There were no atoms of desire, of life, or of soul as there had been for some of the earlier Greek atomists; and hence the new physical science of the seventeenth century had nothing to say about the spiritual or psychological side of life. The physical was set against the mental as a world apart, and in turn the mental came to be seen in terms that were not physical. Two whole sets of opposing categories emerged for describing the two radically different realms of existence, and for the most part they remain with us today, fixed in our way of perceiving ourselves.

Our minds are private, located everywhere and nowhere at once, and impervious to physical measurement. We can't say that the mind is eight inches wide and weighs three pounds, as we can of the brain, nor can we see it or allow others to, as we might an arm or a leg. Our minds are filled with hopes and fears, motivated by desire and expectation, given over to the pursuit of goals; while our bodies, being wholly physical things, behave mechanically, not unlike our motorcars or our water taps.

Our minds are interwoven with memory. Our bodies—apart from skills—are blind to all but the moment. Our minds are holistic and seem to emerge from "somewhere" all of a piece, whereas our bodies are clearly made of separate atoms joined together according to the laws of physics and chemistry, each atom unconcerned with its source and replaceable by another of its sort at any time. "A living human body can therefore be constructed out of a sufficient quantity of anything—books, bricks, gold, peanut butter, a grand piano. The basic constituents just have to be suitably arranged," comments Thomas Nagel.[4] We would not say this of the mind.

The American philosopher Herbert Feigl composed a table of these opposing characteristics that divide the world into the mental and the physical (Table 7.1), and said quite rightly that their apparent irreconcilability lies at the heart of what philosophers call the mind/body problem. Given so much contrast, there is little wonder that dualism holds us all within its spell. The obvious alternatives seem equally unpalatable, or just plain impossible.

Take materialism—for example, the claim that the physical aspects of reality are all there really is and that any mental or spiritual aspects are either entirely dependent on matter for their existence or else don't

TABLE 7.1

Mental	*Physical*
subjective (private)	objective (public)
nonspatial	spatial
qualitative	quantitative
purposive	mechanical
possessing memory	no memory
holistic	atomistic
emergent	compositional
intentional	'blind'; nonintentional

(Table adapted from Feigl's book, *The Mental and The Physical*)

really exist at all. For the materialist, there is no "nonextended thinking substance" (which Descartes claimed the mind to be), nor any angels nor deities nor spirits nor immortal souls. For something to exist, the materialist says, it must be substantial, the substantial is the physical, and the physical is made out of matter, which in turn is made out of atoms. Thus we, the "selves" we perceive ourselves to be, are really just so many atoms briefly drawn together. We *are* our bodies, and our minds a mere reflection of various atomic or neural processes.

The motivation for such materialism is ancient, and springs from a wish both to simplify our account of Nature and to rid mankind of what many have seen as religious superstition and fear. The drive to achieve a unified explanation of all things, including ourselves and our place within the universe, became especially strong with the rise of modern science and the passion to believe that the new physical laws could account for everything that is. But since our usual post-Cartesian understanding of the physical, by definition, excludes the mental, this rush to embrace the scientific outlook through a love affair with material reality has led to a denial of what most of us believe to be the best and most interesting side of human nature. Crude materialism simply can't account for consciousness.

At the opposite extreme from a materialist answer to the mind/body problem, some philosophers—the idealists—have proposed that instead of mind depending on matter for its existence (or not existing at all), it is really the other way around. Mind is primary; mind mediates, and even to a large extent creates, whatever it is that we perceive or mean by matter. Thus for the idealist, the mind is unquestionably real, whereas the body is just so many impressions or ideas within it.

Idealism has taken many forms, from the most extreme sort, which asserts the material world is just a figment of the imagination, to the more cautious sort that argues that matter itself is real enough in some sense, but all the *perceived* qualities of the material world depend upon mind. Some variations on this theme arise from those interpretations of quantum theory that argue either that consciousness collapses the wave function and is thus necessary for the creation of reality, or that it makes no sense to ask whether, or what, matter is beyond what our observations make of it, because those observations are all we can ever know.

But in any form, idealism doesn't sit well with our commonsense intuitions about the world of experience, and it is ill suited to the pursuit of objective science—to wit, the new subjectivism arising from popular quantum physics. It is a theory that satisfies few people who want to understand the relationship between real minds and real bodies.

Because neither materialism nor idealism seems an adequate response to the mind/body problem, there has always been a third traditional way of addressing it, that of panpsychism. If bodies without minds are too brute, and minds without bodies too ethereal, perhaps there is really no way they can be separated after all. Perhaps the mental is really a basic property of the material and vice versa. Perhaps the basic, underlying "stuff" of the universe is just one "thing" that possesses two aspects.

It was clear in our discussion of the possibility that electrons might be conscious that panpsychism of one sort or another has appealed to philosophers and scientists since the beginning of recorded thought. It has colored the thinking of people as widely separate as Parmenides and Heraclitus, Spinoza, Whitehead, and Bohm. Its attraction, like that of materialism, lies in the wish to find one unifying substance that undercuts all divisions of the world into the mental and the material. Unlike materialism or idealism, it tries to do so without denying the reality of either.

A limited panpsychism that associates some sort of very primitive conscious properties with the basic constituents of matter is closest to the argument unfolding in this book. But there are important differences between my brand of panpsychism and the more traditional one.

In the first place, no form of panpsychism so far developed gets to the real heart of the problem. Even if we say that the mind and the body are essentially, in their deepest being, intertwined because all the material constituents of the body themselves possess mental properties, we are still at a loss to say what a mental property *is* and how matter could *have* it. To that extent traditional panpsychism doesn't solve the mind/body problem; it only pushes it back to a more primary level of reality, where ultimately, if electrons *are* conscious, we then have to say that *they* have a mind/body problem.

It is also true, and certainly relevant when trying to make any panpsychist resolution of the mind/body problem sound convincing, that panpsychism in almost any form so far conceived is vaguely embarrassing. It makes people shift in their seats. Even when disclaimers like "very primitive," "elementary," and "proto-" are used to discuss the consciousness of elementary particles, one can't help conjuring up images of electrons falling in love or fretting about whether they might not perform well in their next two-slit experiment.

Such embarrassment causes many who are led to panpsychism for want of any better theory to feel that they must apologize. As Feigl has been quoted saying, "If you give me a couple of martinis, a good dinner and a couple of after-dinner drinks, I would admit that I am strongly tempted towards (a rather watered-down, innocuous) panpsychism."[5]

A few stiff drinks make most problems easier to bear, but seldom solve them. In the cold light of sobriety, one is still left with a notion that, in its present form, jars the modern sensibility. So for that matter do dualism, materialism, and idealism. Something is deeply wrong with all the traditional approaches to the mind/body problem, because ultimately they all rest on outmoded notions of matter or they fail to show how any more updated notions—those following from quantum physics—could do much to explain how anything going on in our very physical (objective) brains might give rise to all the mental characteristics associated with (subjective) mind. The problem has seemed so great that some modern philosophers claim it may have no solution. According to Oxford's Colin McGinn, "Mind may just not be big enough to understand mind."[6]

More optimistically, the problem may simply need a very different sort of approach, one that combines the latest understanding of the physics of matter with what we can surmise about the physics of consciousness.

If we bring together the concept of matter that arises in quantum theory with a quantum mechanical model of consciousness itself, the whole feel of the mind/body relationship changes radically, and does so in a way that illuminates both the true double-sided nature of quantum reality and the meaning of consciousness.

Quantum-level matter, we must remember, is not very "material," certainly not in any sense that would be recognized by Descartes or Newton. In place of the tiny billiard balls moved around by contact or forces, there are what amount to so many patterns of active relationship, electrons and photons, mesons and nucleons that tease us with their elusive double lives as they are now position, now momentum, now particles, now waves, now mass, now energy—and all in response to each other and to the environment.

Existence and relationship are inextricable in the quantum realm, as they are in everyday life. They are the two sides of the quantum coin, and they are essentially what we mean by the wave/particle duality—just as the mind and the body are the two sides of our human existence, or our unfocused, background awareness and concentrated thought are the two sides of our mental life.

The wave/particle duality is a good metaphor for a deeply integrated mind/body relationship, but given the idea that consciousness itself arises out of a coherent ordering of virtual photon relationships in the brain's quantum system (its Bose-Einstein condensate), it becomes much more than a metaphor. *The wave/particle duality of quantum "stuff" becomes the most primary mind/body relationship in the world and the core of all that, at higher levels, we recognize as the mental and physical aspects of life.*

Because it is so primary, and thus irreducible to any other thing or process, the wave/particle duality allows us to see the *origin* of the mental and the physical and to see what we really mean by each.

In any quantum system of two or more particles, each particle has both "thingy-ness" and "relating-ness," the first due to its particle aspect and the second to its wave aspect. It is because of the wave aspect and what it allows to happen that quantum systems display a kind of intimate, definitive relationship among their constitutent members that doesn't exist in classical systems.

If we have, for instance, a group of Newtonian billiard balls bouncing about in a box, they do have a kind of relationship to each other. They bump into one another and alter each other's positions and momenta. They stop each other from occupying the same place at the same time. They attract each other due to the force of gravitation, and if electrically charged might attract or repel each other accordingly. Some of them, if bigger and bouncier than the others, might be said to dominate the smaller and less resilient ones.

Yet these are all *external* relationships. They influence the behavior of the balls, but they don't alter their inner qualities. Regardless of the forces acting on them, they remain round, bouncy, and quite separate billiard balls, each with its own mass, position, and momentum.

But a group of electrons similarly bouncing about in a box will relate in quite a different way. Because electrons are both waves and particles (both at the same time), their wave aspects will interfere with each other; they will overlap and merge, drawing the electrons into an existential relationship where their actual inner qualities—their masses, charges, and spins as well as their positions and momenta—become indistinguishable from the relationship among them. All are affected by the relationship; they cease to be separate things and become parts-of-a-whole. The whole will, as a whole, possess a definite mass, charge, spin, and so on, but it is completely indeterminate which constituent electrons are contributing what to this. Indeed, it is no longer meaningful to talk of the constituent electrons' individual properties, as these continually chop and change to meet the requirements of the whole.

This kind of *internal* relationship exists only in quantum systems, and has been called relational holism.[7]

To get a better feeling for relational holism, let's imagine a more everyday example. Let's say that when we flip two coins at exactly the same moment, the result of the toss will always be one head and one tail. The system as a whole has the property that the coins are negatively correlated. In this example, it's completely indeterminate which way each individual coin falls, but each will always fall in a way opposite to the other. The system, does not, however, *cause* the coins to fall this way or that—it simply draws them into a relationship within which they are negatively correlated.

The pennies are similar to the earlier example (Chapter 2) of the quantum boatmen who were linked across time in such a way that each always used whichever boat the other had not used, and both are everyday analogies with the real proton correlation experiments that initially established the truth of quantum nonlocality.

This kind of quantum relationship, which creates something new by drawing together things that were initially separate and individual, is very important and in itself opens new vistas in the philosophy of physics. But its importance goes far beyond physics.

I believe that such a relationship is both the origin and the meaning of the mental side of life.

In saying this, I am suggesting that consciousness, or the mental, *is,* at the most primary level of existence, a pattern of active relationship, the wave side of the wave/particle duality. Similarly—and this is much easier to understand—the physical side of life originates in the particle side of that duality (Figure 7.1). This essential definition of consciousness as relationship can be applied and found to hold good for all levels and degrees of consciousness.

At the level of consciousness that we understand, that which originates in our own brains, quantum "relational holism" could arise from the correlations of waves in the brain's powerful electromagnetic field created by the jiggling of charged protein or fat molecules in the neuron cell walls. Their relationship would form something like a Fröhlich-style Bose-Einstein condensate,* the most highly ordered form of relationship possible in this world. This state of affairs then gives rise to the unity of consciousness, the "blackboard" on which all our thoughts, feelings, and perceptions are written.

The interesting thing about seeing consciousness in this way is what it tells us something important about humankind's place in the general scheme of things. The relationship between these correlated wave patterns in the human brain and that existing between the correlated wave aspects of two protons or electrons in a simple quantum system are in principle the same. In some important sense, our consciousness is the relationship between elementary quantum particles writ large.

Thus by understanding the quantum mechanical nature of human consciousness—seeing consciousness as a quantum wave phenomenon—we are able to trace the origin of of our mental life right back to its roots in particle physics, just as has always been possible when seeking the origin of our physical being. The mind/body (mind/

*That is, a Bose-Einstein condensate of the kind found in living tissue.

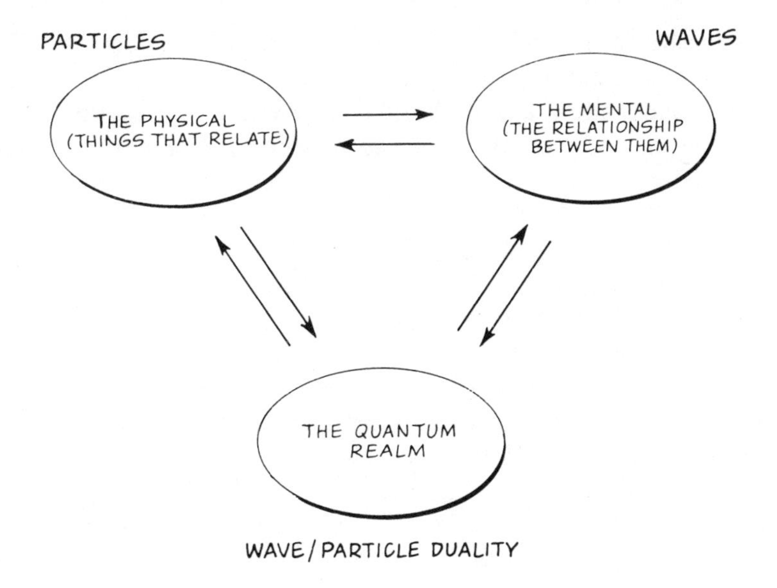

Figure 7.1. The mental and the physical have a common source in quantum reality.

brain) duality in man is a reflection of the wave/particle duality, which underlies all that is. In this way, human being is a microcosm of cosmic being.

We are, in our essential makeup, composed of the same stuff and held together by the same dynamics as those which account for everything else in the universe. And equally—which brings out the enormity of this realization—the universe is made of the same stuff and held together by the same dynamics as those which account for us.

If we interpret consciousness in this way, as a particular kind of creative relationship made possible by quantum wave mechanics, several things fall into place that offer a better understanding of both consciousness itself and its relationship to matter, such as that in our brains.

Most important, if we want to combat materialism and its whole reductionist ethos, this insight allows us to argue that the mind is not some mere offshoot of brain function. Just as the relationship between two electrons whose wave functions are overlapping cannot be reduced to the individual characteristics of the two electrons, so the relationship between the waves that make up the Bose-Einstein condensate of consciousness cannot be reduced to the activities of individual vibrating molecules. We are not our brains.

The condensate is a thing in itself, a new thing with qualities and properties not possessed by its constituent parts. As Plato said in the *Timaeus:*

> Two things alone cannot be satisfactorily united without a third; for there must be some bond between them drawing them together. And of all bonds the best is that which makes itself and the terms it connects a unity in the fullest sense.[8]

He made a similar point in the *Symposium* about two people falling in love, suggesting that there are then not just the lover and the beloved, but also a third thing, which is the love between them. Martin Buber even calls it "the between," the binding force that draws and *I* and a *Thou* into an *I-Thou.*[9]

Love is a particularly apposite example of relational holism, but there are other analogies that help to make the idea familiar.

Think, for instance, of the game of chess. Its "molecules," or "brain matter," are the board and thirty-two pieces, but chess itself is more than these carved pieces of wood. The game is a shifting pattern of rules and relationships, relationships between the pieces and between the players who move them, between their calculations and their psychologies, and it is these that give meaning to the actual mechanics of play.

Or, as an example more useful still because it raises the whole question of art and its meaning, consider the van Gogh painting of the pair of peasant's shoes. The material substrate of the painting is its canvas and the blotches of paint spread around on it, but the work of art that makes us marvel anew each time we look at it can't be reduced to these things, nor to van Gogh's purposes and intentions, nor even to his life history.

The painting is a thing in itself, a whole that reveals something about the world that was never revealed before, and it does so by bringing together (relating) the shoes and the peasant who wore them, his labor and the soil in which that labor was spent, all the things that soil and earth represent for us. In his essay on aesthetics, the German philosopher Martin Heidegger associates such wholeness with the revelation of truth and Being:

> Truth happens in van Gogh's painting. That does not mean that something is rightly portrayed, but rather that in the revelation of the equipmental being of the shoes, that which *is* as a whole attains to unconcealment. . . .[10]

And:

> . . . The essence of unconcealment of what *is* belongs . . . to Being itself.[11]

Relational holism, which is the essence of consciousness (its unity), is also the essence of art and truth. And the bridge between this kind of wholeness and the physical world—hence the bridge between mind, truth, and beauty and the world of matter—can at last be understood when we trace it back to the origins of each in the wave/particle duality. At that most primary level, neither waves nor particles are reducible one to the other. Their existence together is an inextricable unit. As the Roman philosopher Lucretius expressed it:

> For the two are interlocked by common roots and cannot be torn apart without manifest disaster. As easily could the scent be torn out of lumps of incense without destroying their nature as mind and spirit could be abstracted from the whole body without total dissolution. So from their earliest origin the two are charged with a communal life by the intertangled atoms that compose them. . . . It is by the interacting motions of the two combined that the flame of sentience is kindled in our flesh.[12]

Lucretius believed the spirit consisted of "atoms of spirit," and he is thus classed as a materialist in traditional terms, but were his "atoms of spirit" simply translated to "waves of spirit" (relationship), as they might well have been had he been aware of quantum physics and the wave/particle duality, his passionate belief in the subtle unity of mind and body would be very like that being developed here. Perhaps today's materialists might undergo a similiar conversion were they to become more aware of developments in modern physics.

It also follows from the view that consciousness is a kind of quantum relationship that it can in no sense be a "property" of matter, as many panpsychists argue. It cannot be traced back to the being of one elementary particle of matter because it is essentially a relationship between two or more particles. Consciousness is, in its essence, relational, and it can arise only where at least two things come together. "It takes two to tango."

Thus the very most elementary form of mentality possible in this world would be some very primitive consciousness associated with two particles with overlapping wave functions. Anything higher than that, the many states and degrees of consciousness, would be dependent on the many kinds and degrees of relationship—which would themselves, in turn, depend on the many kinds and degrees of structure. Our human consciousness, therefore, is not different in *kind* from that associated with more elementary life forms or with elementary matter, but is different in degree and complexity.

Finally, it is important to ask whether primitive mental properties are associated with all quantum systems where wave functions overlap or only with some particular sort. Is there some special reason why consciousness seems to be vested in living systems, or is this simply a prejudice of ours that blinds us to the mental life of everything around us?

In fact, particles in the world of Nature come in two basic sorts, fermions and bosons. Fermions, which are the particles that combine to give us matter—electrons, protons, and neutrons—are essentially antisocial. Their wave functions can overlap somewhat, but never entirely. They are always to some extent individuals.

Bosons, on the other hand—photons and virtual photons, plus and mins *w* particles, neutral *z* particles, gluons, and gravitons (if they exist)—are particles of relationship. They are the particles that carry the forces that bind together the universe (Table 7.2), and they are essentially gregarious. Their wave functions can overlap to such a degree that they merge totally, thus causing them to share their identities and surrender all claims to individuality.

A Bose-Einstein condensate, which is the extreme degree of relationship required to provide the unity of consciousness, is so called because it is made up of bosons—in the case of a Fröhlich system in living tissue, of virtual photons.

But if the sharing of identity through complete overlap of constituent wave functions is basic to the unity of consciousness, it would seem that only boson pairs (or larger boson systems) would be associated with primitive mentality. Thus what is special about living systems, though

TABLE 7.2

Bosons as Binding 'Particles of Relationship' in Nature

Force	*What It Does*	*Boson*
electromagnetic (the force relevant to everyday life as we perceive it)	binds electrons to atoms; responsible for some chemical bonds; present in all living tissue. Makes consciousness possible?	photons; virtual photons
weak nuclear	binds the nucleus	neutral z, ± w
strong nuclear	binds quarks together to make particles (3 quarks make a proton or a neutron)	gluons
gravitation	binds masses to each other (so holds the universe together)	gravitons?

not unique to them, is their capacity to support Bose-Einstein condensation, and hence mentality.

However, this doesn't offer any really clear-cut division of the material world into some things that are "proto-conscious" and some things that are not, because under some circumstances fermions can form pairs, which then, in combination, behave like bosons.

The covalent chemical bonds that join together organic molecules are in fact built on the properties of such paired electrons. They are also what give superconductors their special properties, just as helium-4 nuclei (paired protons and neutons) are the basis of superfluids. Both superfluids and superconductors, like Fröhlich systems in living tissue, are Bose-Einstein condensates, and probably in combination with the right artificial computing system could generate some structured mentality of a useful sort. This might be a basis for quantum computers.

But under normal conditions, fermions keep to themselves. The relational holism found in fermion systems, which account for most of the everyday material world, is a kind of first cousin to the extreme relational holism of boson systems where particle identity is shared, but there is a crucial distinction. Were this not so, there would be no degree of difference between the world of matter and the forces between matter, nor between the mental and physical realms, and things could not be the way they are.

It is because fermions can't get into the same state (share identities) that matter can be solid. Matter depends on their unsociability. On the other hand, it is because bosons *can* get into the same state that we have large-scale waves and forces. But there is no sharp cutoff point, and in the right circumstances even the most determinedly individualistic fermion can be drawn into a deeper relationship. No recluse is entirely safe from the temptations of society.

In fact, this tension between particles and waves at the quantum level does seem to mirror in an interesting way the similar tension between individuals and groups in human society, raising the whole question of the meaning and nature of individual and group identity as we experience them, and whether the roots of each might lie in the quantum mechanical nature of consciousness.

CHAPTER 8

THE PERSON THAT I AM: QUANTUM IDENTITY

From far, from eve and morning
  And yon twelve-winded sky,
The stuff of life to knit me
  Blew hither: here am I.

—A. E. Housman
"A Shropshire Lad"

Here am I, my body made of elements that once were stardust, drawn from the far corners of the universe to flesh out, however briefly, the pattern that is uniquely me, my soul, a thing that can breathe in the enormity of such awe-inspiring origins. But who or what is this "I" that I think I am?

At this moment, if I direct my attention towards myself, I feel very certain that I exist as a person, that there is something it is right to call "me." This something has its own point of view, its own projects, its own relationships. But am I right to be so certain, or am I perhaps the victim of an illusion, what the Buddhists would call the illusion of self? Is there really such a thing in this world as "me"?

And if I really do exist, how much of me is it right to call "me"? Where do I begin and where do I end?

Am I the same person who was an infant in my mother's arms forty-odd years ago, or the awkward teenage girl who blushed with shame whenever someone pronounced my name? Am I the young woman who married twelve years ago, who had so little understanding of commitment and intimacy and not a clue what it would be like to bear children of my own? Indeed, am I the same person who went to bed last night, surrendering my consciousness and all that I am to the nighttime world of sleep?

Most of us ask questions like these about ourselves or those we know from time to time, usually at moments of personal transition like leaving home or graduating from college or getting married. We answer them as best we can, but with an array of contradictory phrases that reflect an underlying fuzziness in the folk wisdom concerning what it means to be a person.

On the one hand we say things like "I'm not the same person who used to do those things/like those things/talk like that," or "He's only half the person that he was," while on the other hand we say, "I haven't really changed," "I'll always be me," "I'm still the person that I was."

Philosophers also ask such questions, but it is their business to confront the contradictions in what ordinary people say, and when they do so it becomes clear that the existence and identity of persons are very big problems. Given the accepted science of our day, there shouldn't be such things as persons.

For instance, if persons are real, what holds them together? Each of us is an organism made up of billions of cells, with each cell in some sense possessing a life of its own. Within our brains alone, some ten billion neurons contribute to the rich tapestry of our mental life. Another ten billion cells keep the heart beating, the same number again give us a liver, and so on. How, given all this complexity, are we, in sum total, one thing? Or indeed, as some philosophers wonder, is it even true that we are?

The unity of the person, or his supposed unity, poses much the same problem as the more basic unity of consciousness. It raises the same difficulties for any conventional approach, especially if we accept that in all probability our personhood is at least in some part dependent upon the structure and function of our brains. If our brains consist of all those myriad neurons, from whence emerges "a person"? How really solid or basic is he?

The apparent impossibility of answering that question on the basis of known science has been the strength of the dualist case that the mind, or the soul, or the person has an existence of its own that is simply "encased" within or "attached" to the body. But fairly recent research done on the effects of split-brain surgery raises what appear to be insurmountable objections to any theory that tries to separate the person from his brain.

In some cases of very extreme epilepsy, doctors found that relief could be given the patient by severing the bridge (the corpus callosum) between the two halves of his cerebral cortex—that is, by effectively cutting the most human portion of the brain in half. For many years this operation appeared to have no particularly unpleasant side effects, but when people who had undergone it were eventually subjected to tests in a psychological laboratory, the results were dramatic. As one surgeon put it, the brain bisection had effectively resulted in the bisection of personality—where once there had been one sphere of consciousness, now there were two.[1]

Each side of the cerebral cortex, each of the hemispheres as they are called, has its own set of rather specialized functions. The right hemisphere, which for the most part controls the left half of the body, is the more musical, the more intuitive, the center of spatial imagination; the left, which is mainly connected to the right side of the body, is the more logical, the better at calculation, and the exclusive center of speech. In normal brains, the two hemispheres pass information back and forth to one another, thus acting as a coordinated unit, but when they are separated by surgical incision, they lose this ability. It then literally becomes a case of the right hand not knowing what the left hand is doing.

If a split-brain subject (with the right and left visual fields artificially segregated, so that normal eye motions can't help the brain coordinate information) is presented with some objects in his left visual field—the side no longer connected to the speech center in the left hemisphere—he will adamantly deny that he sees anything. If asked to hold two identical objects, one in each hand, he cannot say whether they *are* identical; if asked to pick up an object lying midway between his two hands, he will engage in a tug-of-war with himself as each brain hemisphere struggles independently to carry out the command.

Somehow the creation of two brains results in the effective creation of two selves, each privy to its own sources of information and fired with its own sense of purpose. Perhaps more striking still, these two selves can then again be united to form one coordinated self the moment the experimental constraints are removed. There is enough secondary connection between the two severed hemispheres via the brain stem so that without the artificial separation of their areas of visual stimulation, they can somehow manage a coordinated performance.

The picture that emerges from the split-brain research is of a self that can be divided into two selves and then patched together again under the right circumstances. This person is first one person, then two people, and then again only one. These facts have very great implications for the whole question of personal identity and must make us think again about any notions we might have had about the meaning of our own personhood.

For some philosophers, this is all the proof that is needed to make a case that not only does the self reduce entirely to the mechanics of the brain, but that, to put it stronger still, the existence and continuity of various brain states are all that is meant by the self in the first place. If split-brain cases highlight the fact that in extreme circumstances the self appears to be really just the coordination of two subselves, then in ordinary people the unity that we think of as the mind is only, in the words of Thomas Nagel, "an enumeration of the types of functional integration that typify it."[2] He adds:

> It is possible that the ordinary, simple idea of a single person will come to seem quaint someday, when the complexities of the human control system become clearer and we become less certain that there is anything very important that we are *one* of.[3]

Oxford's Derek Parfit, probably the most important living philosopher writing about the problem of personal identity, pulls even fewer punches than Nagel. He says:

> Given what we now know, what I really am is my brain. On this view, moreover, I am *essentially* my brain. . . . Personal identity is not what matters. Personal identity just involves certain kinds of connectedness and continuity.[4]

The underlying person that I feel myself to be is just a chimera. There is no personal identity, no "deep further fact"[5] that is an abiding me. "I" don't exist.

To some extent Parfit's denial of an abiding personal identity is reminiscent of similar denials by the Continental existentialists, particularly Heidegger and Sartre, whose conclusion that at the heart of the self there is nothingness ("I am the null basis of a nullity"[6]) has contributed to much of the nihilism in modern philosophy.

Parfit himself subscribes to a Buddhist view of life and the self. He feels that his argument against the reality of personal identity, originally inspired by the split-brain research, has liberated him from the prison of self.

> When I believed that my existence was such a further fact, I seemed imprisoned in myself. My life seemed like a glass tunnel, through which I was moving faster every year, and at the end of which there was darkness. When I changed my view, the walls of my glass tunnel disappeared. I now live in the open air.[7]

In feeling this way, Parfit is like those writers on quantum physics—Fritjof Capra and Gary Zukav—who subscribe to a Buddhist view of matter and want to liberate us from the prison of particles. As Zukav argues:

> Photons do not exist by themselves. All that exists by itself is an unbroken wholeness that presents itself to us as webs (more patterns) of relations. Individual entities are idealizations which are correlations made by us. . . . The new physics sounds very much like old eastern mysticism.[8]

But I want to argue that looking too hard for parallels between modern physics and Eastern mysticism distorts our perception of matter, just as being drawn too much towards a Buddhist view of personal identity distorts our perception of the self. Both express a desire to transcend the separateness of things and selves, "an emotional rebellion against the externality of things and . . . a craving to escape the burden of self-consciousness,"[9] as Arthur Lovejoy says in his *Great Chain of Being.* In doing so they leave out a whole side of both reality and human existence.

The time when "all that exists by itself is an unbroken wholeness," when all things are possible and everything is equally real (and unreal), is a time of prebirth, a time before the beginning of time and of a world with a history, with choices and opposites and conflicts. It's the "dream time" of the aborigines, Jung's "uroborus"* or Freud's oceanic feeling,

---

*The ancient symbol of unity and unbroken wholeness depicted as a serpent swallowing its own tail. "I am the alpha and the omega, the beginning and the end." (Erich Neuman, 1954, Part I.)

the time of life in our mother's womb. It is a stage in the development of human consciousness as well as in the development of each individual's consciousness.[10]

But in real life, just as in all the creation myths, there is a moment of birth, a moment when "God separated the light from the darkness," or, in terms of modern physics, a moment when the quantum wave function collapses.

Particles do exist, and so do selves. If they did not, a great many things that we take for granted about our world—the nature of our subject-predicate logic, the whole basis of our morality—would have to be different.

Individuals, of both the particle and the human varieties, make things happen; they bear responsibility. At both extremes of existence, the microscopic and the human, individuals are the focal points of events and differentiations. As my five-year-old daughter says quite sensibly, "If we were all the same, we'd get very confused." If particles were all the same, Nature would get very confused. Indeed, all things considered, if we take a quantum view of the self, the nature of identity in the realm of elementary particles has a great deal to tell us about our own more personal identities, especially about the dynamics by which the self can be "split," as in the split-brain research, and still be a self in some meaningful sense.

Certainly split-brain phenomena are proof that the self is not an eternal, indivisible whole as Descartes argued, any more than particles are the tiny, solid, and indivisible billiard balls that Newton's physics supposed. Both selves and particles are more fluid than that, both a little more "shifty." They flow into and out of existence, now standing alone, now wedding themselves to other selves or particles, now disappearing altogether—teasing us with their dancing forms and shadows.

Less dramatically than having our brains cut in two, most of us have had the common experience of having selves within ourselves, of having pockets of awareness that seem temporarily split off from the mainstream of consciousness, or even whole sides of ourselves that we seldom get a look into.

Pockets of childhood pain rise up to color and influence our reactions to present situations; traumatic memories may suddenly take us over and trap us in some past event. Someone's "conventional" side may dress conventionally, hold down a fairly workaday job, and have a similarly conventional circle of friends, while his "rebellious" side may shun responsibility, favor blue jeans and black shirts, and keep company with an assortment of eccentric, offbeat friends. Any meeting of the two sides can be cause for grave embarrassment.

Psychotherapists are very familiar with these inner dynamics of the self and often get patients to set up conversations between their various subselves in order to bring these more fully into the mainstream of consciousness.* And yet no psychotherapist would argue that because the self is a house with many mansions, it is in any sense less of a thing in itself.

It's just that the self must be defined in some new terms that can take into account its composite nature without denying its substance. What can be the physics of this? And why does the physics of elementary particles tell us more about this self?

In answering these questions we can begin to see for the first time the outlines of the quantum self and why it is such a revolutionary concept.

Like the self, elementary particle systems are wholes within wholes, or "individuals" within "individuals." Because of the wave/particle duality, the constituent members of particle systems carry at all times the properties of both waves and particles. In their particle aspect they have the capacity to be something in particular that can be pinned down, if only briefly and only somewhat. In their wave aspect they have the capacity to relate to other "individuals" through the partial overlapping of their wave functions. Through their relationships, their overlapping wave functions, some of their qualities merge in such a way that a new whole is formed.

The properties of the new "individual" are influenced by those of the "subindividuals" of whose relationship it consists. In every respect, however, it now behaves like a new entity in its own right, with its own wavy aspect and its own capacity for further relationships on its own terms.

This is the notion of relational holism introduced in the earlier discussion of the way our minds and bodies relate. A whole created through quantum relationship is a new thing in itself, greater than the sum of its parts.

---

*These dialogues with different sides of oneself feature particularly in Gestalt therapy, as developed by Fritz Perls (1969).

The process of quantum integration by which new and larger wholes are created is endless. At its furthest extreme every particle in the universe can to some degree be related to every other particle, thus creating the quality of unbroken wholeness that can rightly be ascribed to physical reality. But this unbroken whole has a "grainy" quality that is crucial to the world's being the way it is. It is a whole made up of smaller constituent wholes, each of which to some extent maintains facets of its own identity.*

We recognize these quantum patterns in ourselves, and not surprisingly. If the physical basis of human consciousness is a quantum mechanical system in the brain (our Bose-Einstein condensate), we might expect there to be parallels between the composite nature of particle systems and the similarly composite nature of the human personality. The dynamics of the two are much the same.

Like particle systems, our selves are partially integrated systems of subselves that still from time to time assert their own identities. Their boundaries shift and merge as the boundaries of patterns (excitations) within the Bose-Einstein condensate shift and merge. We are at times more fragmented—more child or adult, more conventional or rebellious, more tormented or at peace—and at times more "together," some better integrated self that binds together the subselves more completely.

The selves within selves of the quantum person undulate and overlap, sometimes more, sometimes less (each is a quantum wave function), and their region of overlap at any one moment accounts for the sense of "I" at that moment. "I" am an ever-present witness to the dialogues between my selves, the highest unity of all my many subunities (Figure 8.1). This is the most basic definition of the self at any given moment—the most highly integrated unity of all my many subunities.

But because of the quantum mechanical nature of consciousness and the relational holism of quantum unities, this shifting, composite "I" is not nothing; it is not an illusion. It can never be reduced to a mere collection of separate selves nor to a collection of separate brain states.

*This is because the material world consists of fermions, those mildly unsociable elementary particles that never entirely merge their wave functions.

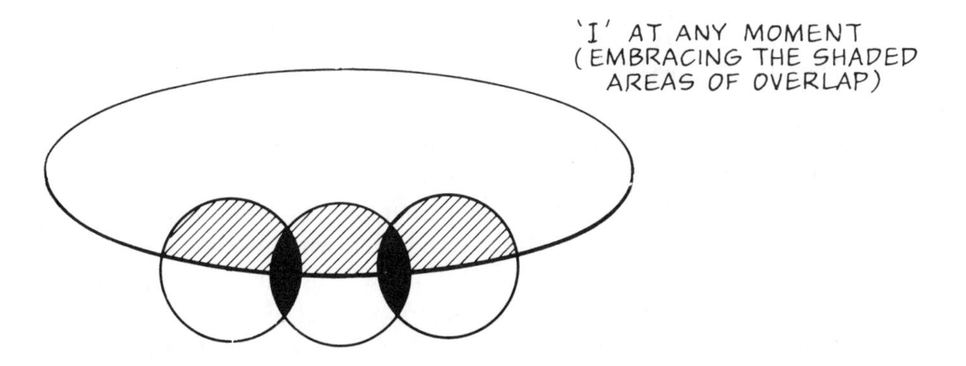

Figure 8.1. The self is the highest unity of many subunities.

"I" am not my rebellious side or my conventional side; both are aspects of me. Nor am "I" the various brain events that give rise to jiggling in the molecules of my neuron cell walls. Quantum systems *can't* be reduced in that way. The unity of the quantum self is a *substantial* unity, a thing in itself that exists in its own right.

And the strength of the self at any moment, the amount of awareness and attention that "I" can bring to bear on my environment or my relationship with others, depends entirely upon the extent to which my subselves (my many pockets of awareness) are integrated at that moment. This is purely a matter of energy and can be understood in terms of the physics of the self.

The Bose-Einstein condensate that gives us the physical basis of consciousness arises from the correlated jiggling of molecules in the neuron cell walls. The extent to which these molecules are correlated, and hence the extent to which the Bose-Einstein condensate is coherent, depends upon the amount of energy pumped into the brain's quantum system at any given moment. If there is less energy available to the system, then the unity of consciousness will be less marked; if there is more energy, there will be greater unity. The range of unity possible in both directions is enormous.

Thus in sleep, for instance, when there is very little energy available to the brain, the self exists in only the most rudimentary and scattered form, if at all, in the form of dream segments. Some people have what are called lucid dreams, during which there is a self there, looking on at the dreams and aware that it is dreaming. This is an instance of there being a higher degree of unity present to consciousness than is common in sleep. In anesthesia, there seems to be no unity at all, and hence no self.

When we are ill, and thus have less mental energy, we have a lower degree of conscious unity than when we are well, and often seem "dull" or "blunted." The self is in a "low key" (our neuron cell-wall molecules are vibrating at a lower amplitude). And even when we are well, the amount of ourselves (our highest unity) that we can muster to deal with the world varies enormously, depending upon whether there are many conflicting inner or outer claims on our attention (pockets of awareness draining off our energy).

People who are in conflict—and this is most of us to some degree—have much less energy available to the main personality (their highest unity) than people who are more integrated. They have many poorly integrated subselves—pockets of childhood pain, pockets of immaturity, pockets of personality that have developed in different directions. Towards one extreme are those who need psychiatric help because they can't "get themselves together," who have so much of their mental energy siphoned off by subselves that they find it difficult to function as a self*. At the other extreme are those charismatic people who sparkle with coherence.

In these terms, coherence has a very physical meaning. A coherent personality is one that rests on a coherent quantum system in the brain. Just as laser beams are brighter than ordinary light because they are more coherent (they, too, are Bose-Einstein condensates), some charismatic people are more radiant than others for the same reason (Figure 8.2). Among those we know, we notice that people weighed down with problems and conflict (who hence carry the burden of scattered attention) seem "dull" for a time; when they resolve their crises, they seem "brighter."

Because quantum systems are always undulating, their boundaries shifting and changing, the extent to which the self is integrated at any one time may change from moment to moment. The act of paying attention focuses our mental energy, so through the mechanism of selective attention we can channel more energy into some particular

*On this interpretation, schizophrenia is an illness based on a problem of energy distribution in the brain.

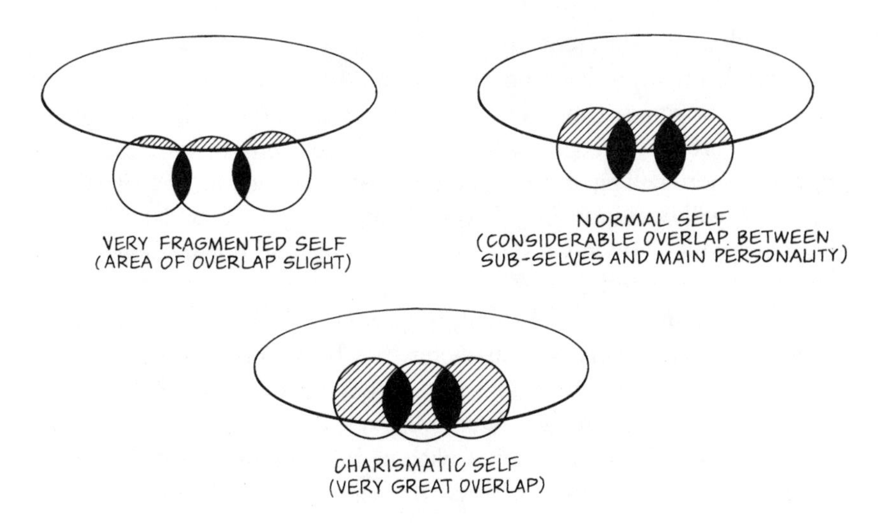

Figure 8.2. The area of highest unity (the extent to which the wave functions of subselves overlap) determines the strength of the self at any moment.

aspect of the self, thus "lighting it up" (giving it more coherence) while others recede into the background. We may even at times be "taken over" by one of our subselves—as for instance when an angry person can think of *nothing* good about the person he loves during a row, or a depressed person can think of *no* reason to be happy when suffering his affliction. When this happens we say the person is "unbalanced," an apt description given the quantum dynamics of the personality.

The quantum self, then, the "I" that we take ourselves to be, is real enough, but from moment to moment it is a shifty thing with fuzzy and fluctuating boundaries. We can talk about its dynamics, but we can't really pin it down, any more than we can pin down both the position and the momentum of an elementary particle. It has substance, but in many important ways that substance eludes us. I can say with some certainty *that* I am, but if this were all there was to the self, it would be difficult to say *who* or *what* I am.

Moment-by-moment fluctuating existence *is* all there is to the individual identities of elementary particles. We can at any one moment say various things about an individual electron (its charge, its mass, its spin) and we can at that same moment tell one electron from another (they will be in different places, if nothing else, or have different momenta), but they have no abiding identity, no identity that stays with them through time. They are very much here today and gone tomorrow. If two recognizably individual electrons merge and then separate, there will again be two individuals, but individuals without a history. It will be impossible, indeed meaningless, to ask which one was originally which. In this respect human beings are not like electrons, or so it seems to most of us.

Electrons, like other elementary particles, are extremely simple things. They have very few characteristics by which they can be distinguished, and most important of all, they have no memories. That is why they have no history. Memory is a means by which we record where we have been and carry it with us into the future. Without it, there seems to be no link between the selves we were and the selves we are now.

At the level of simple common sense, I know that I am the person who went to sleep in my bed last night and woke up there this morning because I remember myself and most of my recent activities. I remember my name, much of my history, my physical appearance, and the fact that I laid myself down to sleep in that bed the night before. Similarly I remember that I often went fishing with my grandfather as a child, that I attended a certain elementary school in Toledo, Ohio, that I studied physics at MIT, that I had a particular circle of friends in Jerusalem, and so on. Through these memories I have a picture of myself as a person who has been in existence across the many divisions of time, a particular person with my own particular history.

But how reliable *is* memory, and just how substantive? Is the history that it gives us a real thing, an aspect of an abiding personal identity that does indeed survive across time, or is it more an illusion, a trick that makes us believe there is some definitive relationship between the selves we were, the selves we are, and the selves we will be, where in fact there is none?

For a philosopher like Derek Parfit, any notion that the thread of memory has substance is a mistake. Being a committed reductionist, he equates the self, such as it is, with the features and dynamics of our current brain states: "I am my brain." As the brain is changing moment by moment, dying, developing, exchanging old atoms for new, so the self is changing—not growing, but in a literal sense becoming different.

Parfit sees a human history as a chain of successive selves, each linked contingently by a degree of "physical and psychological connectedness." The physical connectedness, supplied by the close neural links between one brain state and another, is necessarily short-lived, as the very atoms of which the brain is made change over. The psychological connectedness, supplied by memory, is equally at the mercy of a fleeting thing. Diminish or destroy the memory and we diminish or destroy the connectedness between selves.

From this point of view, memory and the self are separate and each successive self is separate from all preceding and succeeding ones. Speaking of his own present and future selves, Parfit remarks, "If I say, 'It will not be me, but one of my future selves,' I do not imply that I will be that future self. He is one of my later selves, and I am one of his earlier selves. *There is no underlying person who we both are.*"[11] (Emphasis mine.)

Given the image of memory as a thread linking successive selves, a facility of the brain by which it records today's brain states and plays them back to us tomorrow, it would seem that little of substance joins the self across time. But this whole view of the self and memory is too disjointed and Newtonian, picturing successive selves as so many discrete particles being shot along the cannon of time (Figure 8.3). It is the only view available to those who think of the link between the self and its brain in classical (and hence reductionist) terms, but with a quantum view of the self, and a quantum understanding of memory, the whole picture changes radically.

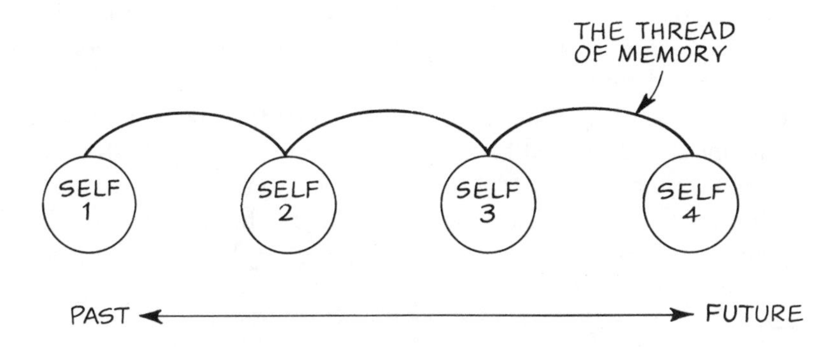

Figure 8.3. Successive, Newtonian selves linked by memory.

To discuss the quantum self and its necessary relationship to what I shall call quantum memory, it might be good to recall the diagram showing how the self emerges from all the many kinds of information fed into the ground state of consciousness (our Bose-Einstein condensate) (Figure 8.4).

The portion of the illustration labeled "present state of consciousness" represents the self at any moment, the self as it is "now." According to psychologists, "now" (William James's "specious present") is a span of time lasting for anything up to twelve seconds, and represents the breadth of experience that our awareness can digest as a unified whole.*

For a quantum self, "now" is a composite of already existing (but ever fluctuating) subselves—our selves as we were *before* "now"—and various inputs from the external world (new experiences), each of which forms its own wave pattern on the ground state of consciousness, the Bose-Einstein condensate. Personal identity on a moment-by-moment basis is formed by the *overlapping* wave functions of all these things, which cause ripples and patterns to appear on the condensate—our thoughts, emotions, memories, and sensations.

As "now" fades into the past, the self that I was then is recorded in the brain's conventional memory system as "a memory of the past." It becomes a new set of neuron pathways, which in turn can feed patterns of energy back into the condensate. This is the familiar sense of memory, the kind spoken of by Parfit and other philosophers. But on a quantum view, the self I was a moment ago is also woven into the next "now," into my future self, by the overlapping of *its* own wave function with all the new wave functions just appearing as the result of new experience. In quantum physics, particle systems can overlap in both space and time.

Thus each self that I was, moment by moment, is taken up into the next moment and wedded to all that is to come—wedded both to old memories, in the conventional sense of memory, as these are fed back into the condensate, and to new experiences. The dynamics of this ongoing dialogue between past and present are very like those by which

---

*Significantly, I think, the "specious present" is roughly equivalent to the "coherence time" of a Bose-Einstein condensate (such as a laser beam)—the length of time during which the system can interfere with itself and sustain a phase relationship. This is another piece of evidence linking consciousness to the physics of Bose-Einstein condensates.

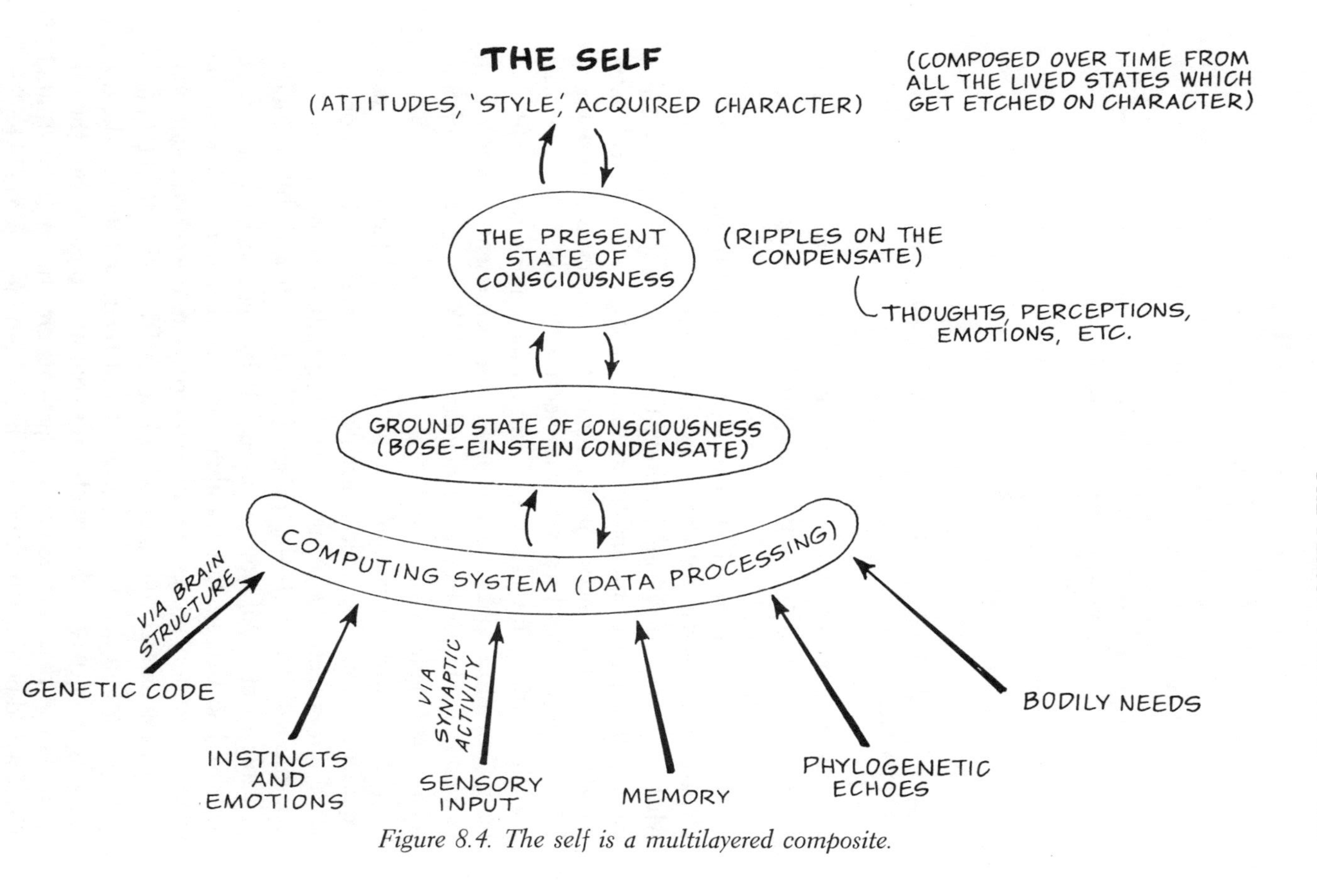

Figure 8.4. The self is a multilayered composite.

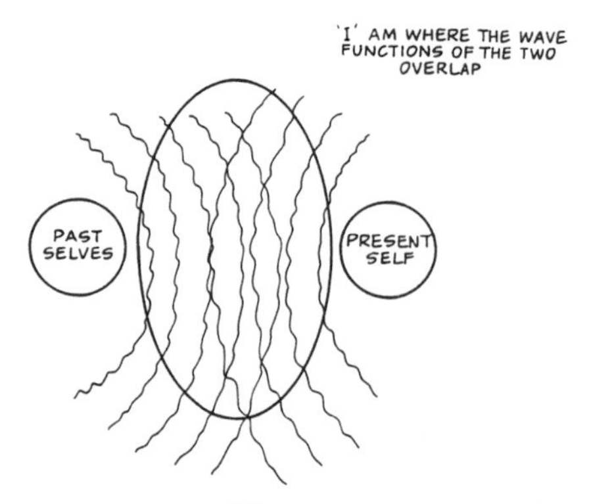

Figure 8.5. Quantum memory: The past gets into a phase relationship with the present.

the wave functions of two elementary particles overlap to form a new quantum system, only in this case what is being formed is a new quantum self.

This weaving of the self moment by moment, as the wave functions of past selves overlap with the wave functions of the present self, is what I mean by quantum memory. It is a necessary, definitive link among our past, present, and future selves, and gives us the mechanism by which we have a personal identity that abides across time. I am, in part, the person that I was yesterday because that person is now woven into the fabric of my being. In quantum mechanical terms, the past has gotten into a "phase relationship" with the present—because both past and present produce wave functions on the ground state of consciousness (Figure 8.5).

Quantum memory is more than just a memory of facts or images or experiences. Someone could forget all these, forget his entire history (as some unfortunate people do), and his quantum memory, his *lived* dialogue with the past—his personal identity—would still be intact, available to others if not to himself. If conventional memory were destroyed so that he lost his ability to record new experiences, the quantum memory process would be interrupted. It works through conventional memories (our past selves) being fed back into the brain's quantum system (Figure 8.6), and without their contribution the dialogue between past and present can't continue to grow. But the self would not in that case be entirely lost—its growth would be stunted and it would become stuck in time, but something of it would be left.

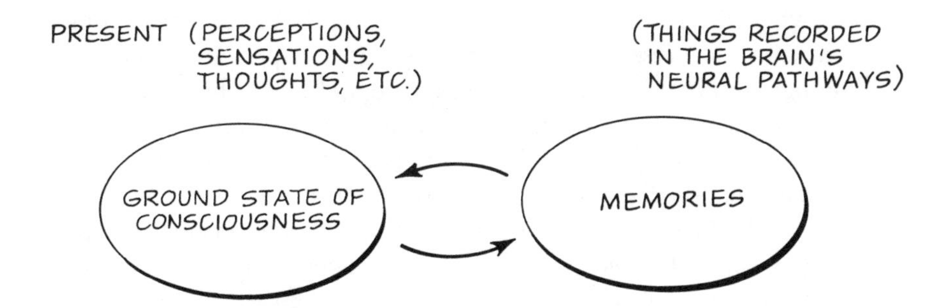

Figure 8.6. Conventional memories are fed back into the brain's quantum system.

Psychotherapists are very familiar with the process of quantum memory, though most might be surprised upon first being told that! The physics by which the wave functions of our various past subselves can overlap with, and hence get integrated into, the present self is the physics by which psychotherapists get their patients to relive past experiences in the "now," thus robbing them of their isolation and their sting and wedding them to the present. This moment of psychoanalytic "insight," during which the past *is,* now, and both past and present are transformed, is quite different from a simple, intellectual remembering of past events.*

Put in quantum terms, the wave function of a relived past moment overlaps with the wave function of now, and the two unite to form a new way forward. The person gains perspective and becomes more coherent.

Thus through quantum memory we take up the past and make it ours in the present. We reincarnate the past (all our past selves), giving it new life in a new form. This is the physical basis of salvation, and of creativity. It also gives new insight into the physical efficacy of ritual (traditional meals like Christmas turkey and Easter lamb, religious rituals like communion and removing the Torah from the Ark, birthday and remembrance rituals, and so on). In observing rituals, we draw the

---

*Both Nietzsche and Heidegger speak of a "moment of vision" during which the past, present, and future are all united in one creative instant.

past, which is organized through the ritual, and all our past observations of it, into the present and integrate it with our present experience. Through observing the ritual we re-create and relive the past.

By reliving past moments, the quantum self is creative on two fronts—on the one hand it reincarnates the past, giving it renewed life and meaning; on the other it re-creates itself at every moment.

Derek Parfit is right to make us see that abiding personal identity is impossible in the old Cartesian, or Newtonian, terms. Given what we now know about brains and the dependence of the self upon them, there is no way we could adhere to a notion of the self as a fixed, definite, and indivisible thing that abides unchanged through time. But in giving up that Newtonian self, we are not left with *no* self.

The quantum self is simply a more fluid self, changing and evolving at every moment, now separating into subselves, now reuniting into a larger self. It ebbs and flows, but is always in some sense being itself. I am the person who was an infant in my mother's arms, who was a teenager, a young woman, and so on, but each of these past aspects of my being was also me as I am now. My relived past can no more be separated from my present than my present can be separated from my past. As T. S. Eliot says, "Time past and time future are both present in time now."[12]

This is the individual, or particle, aspect of the person who I am, the part of my identity that is in dialogue with myself across time. But I am also a person who relates to others, and within a quantum interpretation of the self those relationships also define my identity. Thus to fully know the person who I am, I must understand the relationships that I am—the wave aspect of my being.

CHAPTER 9

THE RELATIONSHIPS THAT I AM: QUANTUM INTIMACY

After when they disentwine
You from me and yours from mine,
Neither can be certain who
Was that I whose mine was you.
To the act again they go
More completely not to know.

—ROBERT GRAVES
"The Thieves"

To know oneself and not to know oneself, to be oneself and yet to escape oneself, to be independent and self-contained and yet to join with others and feel part of something larger than oneself—these are tensions with which we are all familiar.

There are moments when the burden of the self, its perceptions, its responsibilities, its isolation, are almost more than we can bear, and other moments when we will fight with all we have to preserve it, to maintain our sense of "I-ness" and "mine-ness."

In Graves's poem the lovers gladly "lose" themselves in one another, willingly surrender the boundaries that might otherwise have guarded and defined their separate selves. Neither can tell where one ends and the other begins, "who was that I whose mine was you." We all treasure such moments of intimacy, indeed will often risk all to have them, and yet we also strain to break free. Graves calls his poem "The Thieves," implying that both have stolen from each other something that was not rightly theirs to take.

Freud spoke of these tensions between the "I" and the "not-I" in terms of the sex instinct and the ego instinct, the constant battle within the psyche between the drive to merge and the equally strong drive to stay separate.[1] In philosophy the same struggle is reflected in the tension between those philosophers who argue that the individual is *everything* and the world outside him unimportant (perhaps even nonexistent), and other philosophers who want to say the individual is *nothing* and it is only his relationships to other things and people that matter.

The first is the philosophy of radical individualism, which, if taken to extremes, becomes solipsism. The second finds expression in the philosophy of Parfit, in Eastern mysticism, and in Marxism's preoccupation with "history," with society and social forces.

But how real is this tension? Is it a tension between reality and illusion—between a real self and illusory intimacies, or an illusory self and real intimacy—or is it more a tension between two realities, between a real self that exists as an individual and a real "we" of whom that individual can be a part? If the latter is true, if both "I" and "we" are real and important, why do our philosophers have such trouble seeing it?

The whole argument of the last chapter led us to see that the reality of "I" is not a problem. Understood quantum mechanically, the self is a fluctuating and fuzzy thing whose boundaries, both internal and external, are always shifting and changing. It is nonetheless a real thing, a substantial thing. The self is not an illusion.

To common sense, close relationships between ourselves and others seem equally real, and indeed seem at least in part to define who and what we are.

Most of us have had an experience like that of Graves's lovers in which the intimacy between ourselves and another has been so complete as to seem to erase all distinction between us. It is the common relationship between mothers and their babies, where the mother, at least, feels the baby to be an extension of herself and experiences the two of them as existing in a sphere of intimacy whose boundary defines their common identity. Psychologists tell us the baby feels the same.

The same sort of intimate bond is said to exist between a psychotherapist and his patient, where often the therapist finds himself feeling emotions or thinking thoughts that are really those of the patient. The two seem at moments during their fifty-minute hour to share a common identity, to be of one body and one mind. The mechanism by which this happens is called projective identification and is thought to be a crucial vehicle through which the therapist can really know at first hand the unconscious problems confronting his patient. For a time, until he becomes aware of them and their source, he experiences the patient's problems as his own.

As one Jungian analyst describes it, "Projective identification may be conceived as a kind of fusion which involves the mixing and muddling up of subject and object, of inner world and outer world; it involves the undoing of boundaries."[2]

Many instances of this extreme degree of intimacy occur in everyday life. The experiences of Graves's lovers and the mother and her baby are of this sort. So is the bond between a gifted teacher and his pupil, where not just the teacher's knowledge but his whole person—his passion, his mannerisms, his style of thought—"get inside" the pupil and become the pupil's own. Similarly, gifted political leaders have a way of sensing the unexpressed desires and aspirations of their followers and not just *expressing* them as their own but actually *feeling* them as their own.

In all these cases the intimate relationship seems to result in two people's overlapping to such an extent that each takes on the other's inner content. They share an identity. The mechanism by which this happens also seems very closely related to the slightly less extreme sense of common empathy we feel with others. In empathy, we know that we are not the other person but we also know what it would be like to be him, to be in his place, feeling his emotions.

Empathy is one form of intimacy we can share with total strangers as well as those very close to us. There are others as well.

Every day we experience small and fleeting intimacies with other people—the nod of recognition exchanged with a stranger on a country lane, the smile of common amusement shared with the person next to us when a child embarrasses his mother with some tactless comment about a fat lady or a bald man—brief moments when even the company of strangers somehow touches our being, gets inside the boundaries of ourselves, and in some small way leaves an impression there. We are never quite the same again, and neither is the other.

Many things about our experience influence and change us. Most obviously, the health and functioning of our bodies, including our brains, depend upon the quality of the food that we eat and the many shifting factors in our external environment. Similarly our selves, our thoughts, and our behavior are influenced continuously by the thoughts and behavior of others, by members of our families, by friends and colleagues. We are also influenced by the culture at large—by the books that we read, the films that we see, the music we listen to, and so on. Much of what we think ourselves to be depends upon the overall context of our being, and to a large extent these influences are not mysterious. No new theory of the person is required to explain them.

But intimacy seems to be different. Intimacy is not a relationship between "I" and "you," or between "I" and "it," in which you or it (a book, a stone, a computer) influences *me.* In intimacy, I and you appear to influence *each other;* we seem to "get inside" each other and change each other *from within* in such a way that "I" and "you" become a "we." This "we" that we experience is not just "I *and* you"; it is a new thing in itself, a new unity.

The "we" both alters the I and the you who make it up, and takes on its own identity with its own capacity for further relationships.

The "we" that appears to arise in an intimate relationship is the *I-Thou* written about by Martin Buber in his distinction between our relationships with things and our relationships with other people. With things, he says, we have an *I-It* relationship. *It* may influence me, but *I* do not influence *It.*

> The world [of *Its*] has no part in the experience [of my meeting it]. It permits itself to be experienced, but it has no concern in the matter. For it does nothing to the experience, and the experience does nothing to it.[3]

But "when *Thou* is spoken, the speaker . . . takes his stand in relation."[4] In *I-Thou,* I and you become "we."

Because each of us has personal relationships of one sort or another and most of us are fortunate enough to have experienced some intimacy in our lives, the existence of "we" is a *lived* truth for us, but has it any factual basis? Is there really such a thing as intimacy, or is it just an illusion clung to by the isolated self? It seems real enough to us, but when we try to understand it, to express it and explain it and make it part of the conceptual apparatus with which we structure our world, we find there is a problem.

How *can* two individuals meet in such a way that each is changed internally by that meeting and in such a way that the *meeting* itself assumes an identity? What is the physical basis of this "meeting," of this new "we," of Buber's *I-Thou*?

In any classical approach to the philosophy and psychology of persons, this question has no answer, and thus any explanation for how close relationships of the sort obvious to common sense can exist is difficult, if not impossible.

In Cartesian philosophy, which has so deeply influenced the individualist strain of modern thought about the self and its relationships, there are no intimate personal relations. Everything in Descartes's thinking takes place from the first-person point of view of the isolated *cogito,* the "I" who thinks and is nothing but his thinking. Any relationships this *cogito* may have to anything or anyone else it has indirectly, mediated through Newtonian matter or through the mind of God. "I" and "you" never meet.

Cartesian isolation was further underpinned by Newton's physics, where the concept of matter as consisting of so many separate and indivisible billiard balls complemented Descartes's separate and indivisible minds. The notion of a relationship as a set of *external* influences enacted between strangers became the paradigm for all relationships.

Billiard balls don't "meet"; they don't get inside each other and alter each other's internal qualities. They've no means to do so, because each is always and only itself and wholly impenetrable to any outside influence. Like Descartes's minds, they relate to each other only indirectly, by way of external forces that cause them to attract or repel one another or to bump against one another from time to time. During a collision they suffer an impact and may undergo a change of position and momentum, but they remain the same in themselves before, during, and after the collision. Their relationship during the collision, the attraction or the repulsion, is what Sartre would call a contingent truth.

Indeed the whole Cartesian-Newtonian paradigm of isolated individuals having only contingent, external relationships underlies the core of existentialist thinking about interpersonal relations.

In *Being and Time,* Heidegger tells us that *Dasein* (human being) can have no involvements. "When *Dasein* is absorbed in the world of its concern—at the same time, in its Being-with towards Others—it is not itself."[5]

And Sartre, who sees himself as taking the Cartesian revolution to its fullest conclusion, argues in *Being and Nothingness* that the being of others is a fact of our existence, but not an *essential* fact. It's just what he calls a factual necessity. "Being-for-others is not an ontological structure for the For-itself. We cannot think of deriving being-for-others from a being-for-itself as one would derive a consequence from a principle."[6] Our relationships with others are just things that happen, like flies alighting upon our noses. The other does not really get to us. If we think he does, we are in "bad faith."

Freudian psychoanalysis, too, largely influenced by Descartes and Newton and in turn so responsible for the way so many ordinary people see themselves, has no conceptual framework for interpersonal relationships, indeed does not even consider such relationships its proper business. As the author of *A Critical Dictionary of Psychoanalysis* puts it, "This is because psychoanalysis is a psychology of the individual and therefore discusses objects and relationships only from the point of view of a single subject."[7]

According to Freud, it is not *others* who influence us but our own *ideas* about those others, our projections. Freudian influence is always a one-way transaction, what Buber would call an *I-It* relationship, where the other is an object. We are able to take only a representation of this object into our psyches, where we make of it what we will. There is no dynamics of interpersonal relationship, only a dynamics of the individual psyche.

Following this model of relationship as object representation, Freud's pupil Melanie Klein interpreted projective identification, where two people *seem* to get inside each other and share an identity—Graves's lovers, the mother and her child, any *I-Thou* relationship—as really a process by which one is "ingested" into the other as an object of his own fantasies:

> The ego takes possession by projection of an external object—first of all the mother—and makes it into an extension of the self. The object becomes to some extent a representation of the ego, and these processes are, in my view, the basis for identification by projection, or "projective identification." . . . The vampire-like sucking, the scooping out of the

> breast develop in the infant's fantasy into making his way into the breast and further into the mother's body.[8]

Klein, like Freud, Sartre, and Heidegger, has no model for a genuine two-way relationship of the sort that leads to intimacy. None can discriminate between the way we relate to other people and the way we might relate to a machine, because for them all, both machines and people share the quality of being *objects.* All live in the shadow of Descartes's isolated *cogito* and Newton's impenetrable billiard balls, and the work of each is in its own way an inevitable development from those prototypes of detachment.

Thus "bad faith," "object representation," and "vampire-like sucking" are the models of relationship offered to us by some of the most influential thinkers of this century. Each has made its way into the general culture and contributed in no small measure to the sense of alienation felt by so many. Little wonder that other thinkers—Parfit, Capra, Zukav, Bohm—have attempted to transcend this alienation by denying the existence of the isolated and isolating self altogether.

But the self does exist and so, we know in our deepest intuitions, do close relationships. The "I" and the "we" are a case not of either/or, but of both/and. I am uniquely me, something in myself that only I can be, and I am also my relationships with others, something larger than myself.

To transcend this tension between the I and the not-I, we need to ground the reality of "we" in a new conceptual structure that gives equal weight to individuals and to their relationships, a structure that rests on the physics of consciousness. We need to see how it is, physically, that "we" can be both a compound of "I" and "you" *and* a new thing in itself with its own qualities. Such composite individuals are not possible in classical physics, but we know that they are the norm in quantum physics.

This new conceptual structure for interpersonal relations can be found in the tensions within the wave/particle duality and the ability of an elementary particle to be both a wave and a particle simultaneously.

The particle aspect of quantum matter gives rise to individuals, to things that, however briefly, can be somewhat pinned down and assigned an identity. The wave aspect gives rise to relationships between these individuals and the consequent birth of new individuals through the entanglement of their constituents' wave functions. Because wave functions can overlap and become entangled, quantum systems can "get inside" each other and form a creative, internal relationship of a sort not possible with Newtonian billiard balls. Quantum systems "meet," and through their meetings evolve (Figure 9.1).

If there were only the particle aspect of elementary matter, the world as we know it would never change fundamentally. Existing particles would move around and sometimes form new combinations, but the intrinsic nature of matter would stay the same. The world would be noncreative. It is only through the wave aspect and the creation of new individuals to which it leads that the universe evolves.

The tension between particles and waves within the wave/particle duality is a tension between being and becoming. Similarly the tension within ourselves between the I and the not-I, between keeping ourselves to ourselves and engaging in more or less intimate relationships, is a tension between staying as we are and becoming something new. The key to both is quantum wave mechanics.

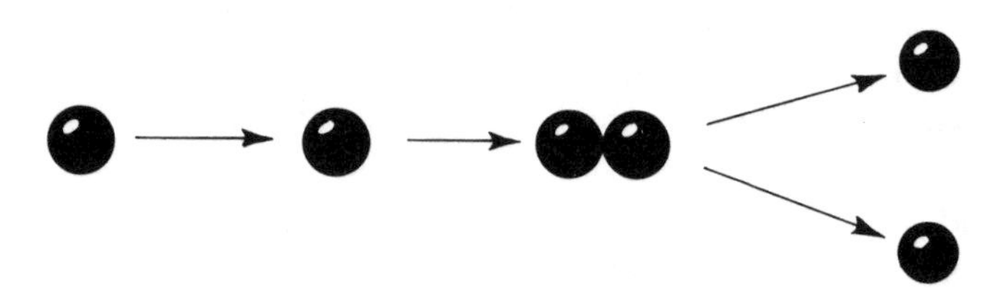

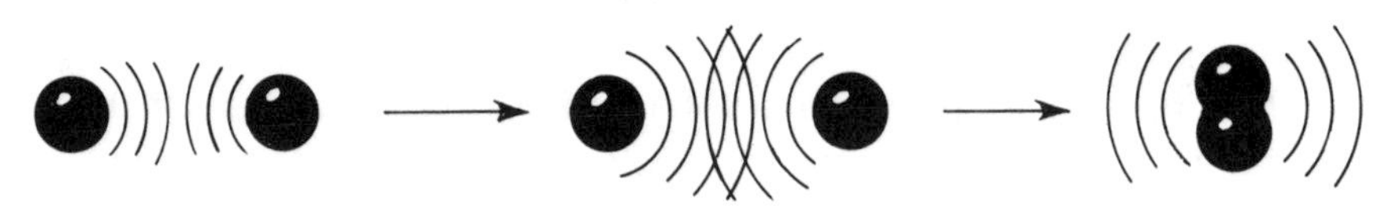

Figure 9.1. Newtonian billiard balls have only external relationships. After a collision they go their separate ways.

Like elementary particle systems, we, too—our personalities, our selves—are quantum systems. Within any one individual, the physics of overlapping subselves can easily be seen as the overlapping of wave patterns within the Bose-Einstein condensate of consciousness. Each of us as a person is a composite of quantum subselves that are also one self (one highest unity).*

As we experience it, the relationship between you and me when we become "we" is very like the relationship binding the many subselves of my own self. It presents the same challenge for interpersonal relations as did the integration of those subselves for personal identity.

We can see this when we compare inner dialogues between our own subselves with dialogues between ourselves and others. I could have much the same conversation between my own rebellious and conventional sides about responsibility as I might have with my daughter's principal about getting the child to school on time each morning.

In principle, the quantum wave mechanics of overlapping persons should be the same as that of overlapping selves within the self, though the actual physics of this is less clear.

Within any one self, we are speaking of overlapping wave patterns on a given Bose-Einstein condensate, whereas between people we are speaking of overlapping wave patterns on *different* Bose-Einstein condensates. Between people, the overlapping effects may be nonlocal, as when two physically separate laser beams interfere with each other across time.[9] But, however they happen, the uncanny way in which quantum wave mechanics fits what we know about interpersonal relationships suggests there is definitely a physical basis of some sort there.

If I can be a self in the first place, a self-for-myself, I can be a self with and for others. Indeed, in any quantum model of the person, it is impossible that it should be otherwise.

When a newborn baby begins his life, his person is mainly genetic. He comes into the world with a rudimentary self but wholly without experience,† and in the early months of his life he is fused with his mother. In quantum terms, his own wave function is almost totally overlapped with that of his mother and they are in a relationship of projective identification.

---

*As discussed in Chapter 8.

†Save whatever nascent experience may be possible within the womb.

To a very large extent, the baby's experience *is* the mother's experience and he begins to weave his self using the mother's cloth. He takes in the mother's responses to the wider world, her perceptions, her emotions, her cares, and lays them down in his own quantum memory system. They become the stuff of which he is made and influence the development of neural pathways in his own brain (Figure 9.2).

This state of fusion with the mother through projective identification is equivalent to Erik Erikson's first stage of psychic development.[10] The physics of it, the physics of primal bonding between mother and baby, is very like the physics of covalent bonding in chemical compounds, where the compound is formed by the two constituent atoms sharing an electron ring (sharing a state).

All his life the baby will carry his mother within him as part of himself, as once she carried him in her womb, and the success or failure of his initial bonding with her will lead to the "basic trust" or "basic mistrust" that Erikson associates with this stage.

As the baby matures, both his self and his perceptions develop to take in things and people beyond the mother. With his senses he adds information about the physical world to his own instinct patterns and

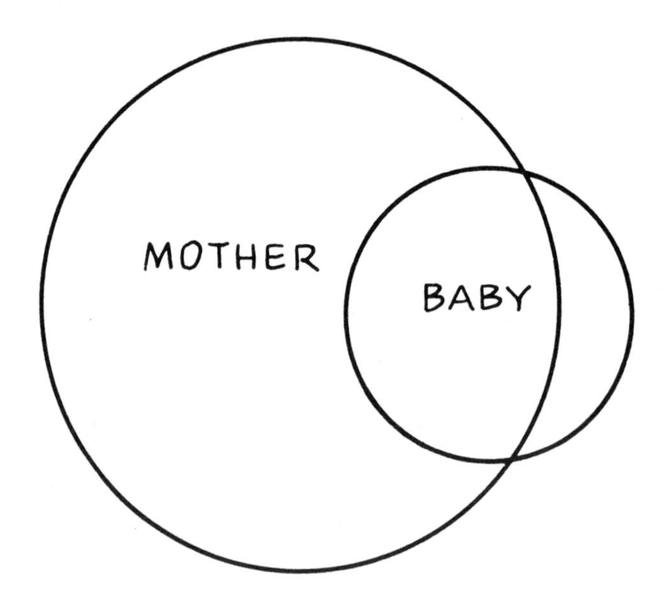

Figure 9.2. At the earliest stage of development, the baby is nearly fused with the mother. This is Erikson's first stage.

builds up a repertoire of reactions to his environment; with his developing self he forms a network of fledgling relationships with others around him. His own tiny wave pattern integrates with those of the people who hold him and engage his attention, and these, too, he weaves into his person.

As the strength and complexity of the baby's own person grows through relationships with others than the mother, the extent of himself (his own wave pattern) entangled with her (her wave pattern) diminishes.

Through the combination of his own genetic material and the relationships that only he can make in the particular way that he makes them (his individuality), the baby begins to feel the tension that pulls us all back and forth between being ourselves and being with others. To assert himself, to allow more of himself to be free to engage in relationships with others, he splits from the mother. Less of his energy is involved in the integration of his wave function with hers and more in the integration with others less likely to overwhelm him (Figure 9.3). This is equivalent to Erikson's second stage, the stage of autonomy versus shame and doubt. In this phase of his life, the splitting phase, the baby or young child fights deep intimacy. His task is the integration of many relationships without losing himself in any one.

Later still, when the child has established himself in his own right, when he has woven enough of a self both to have an inner complexity and to be able to balance that complexity, he returns to an intimate relationship with his mother. In other words, he has grown a Bose-Einstein condensate large enough to sustain several different wave

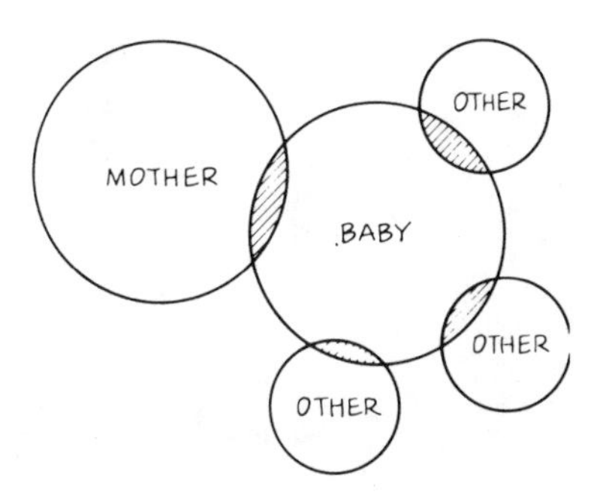

Figure 9.3. The baby splits from the mother to engage in other relationships. Erikson's second stage.

patterns and laid down neural pathways complex enough to make those patterns.[11] He allows part of himself to be connected with her (overlapped with her), safe in the feeling that other parts of himself are free for other relationships (Figure 9.4). This is Erikson's third stage, initiative versus guilt.

The pattern of fuse-split-connect laid down between mother and baby stays with us all our lives, to be repeated time and again to some extent with each new intimate relationship we form and yet again within any one intimate relationship as that relationship evolves. Through the initial fusion the self becomes one with another self, during the split each fights to regain his lost individuality, and in the connect phase each realizes himself within a new combined reality larger than himself.

As Erikson's dual categories—basic trust versus basic mistrust, autonomy versus shame and doubt, initiative versus guilt—show, trouble between mother and baby at any stage in this pattern can cause the baby to remain fixated for life on the negative side of that stage. The path to growth beyond such a fixation is possible through the quantum memory mechanism discussed in the last chapter, where the past is taken up and wedded to the present to form a new reality, and it is through this mechanism that a psychotherapist would seek to help such a fixated person.

Because we possess the quantum memory facility, we need never be stuck, or lost, at any stage of our history. Personal salvation is always possible.

The fuse and split stages of the process of psychic growth in human

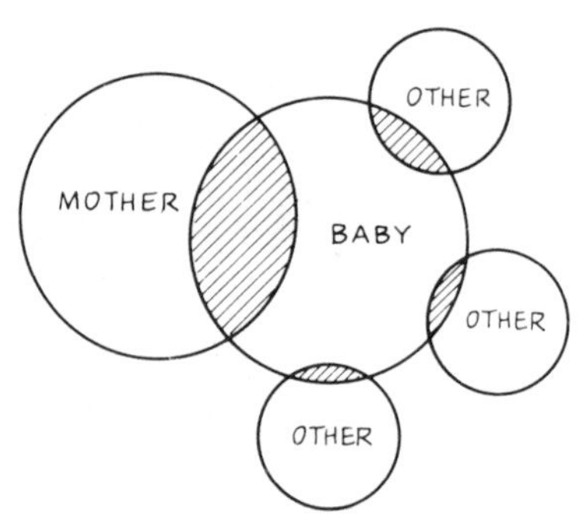

Figure 9.4. The baby comes back to an intimate relationship with the mother in the "connect" phase of relationship. Erikson's third stage.

beings are the same as those for elementary quantum systems as they combine and recombine in larger systems: Each subsystem maintains some identity through its particle aspect and merges with a new and larger identity through its wave aspect. Both we and elementary particles change identities as a consequence of relationships, but we, unlike they, remain changed and accumulate change (accumulate character) because we have memories. Thus only we (or at least only systems complex enough to possess memory) can have a connect phase of growth.

Through the process of quantum memory, each of us carries within himself, woven into the fabric of his own soul, all the intimate relationships he has ever had, just as each of us weaves into his being all of his other interactions with the outside world.

Intimate relationship itself is accounted for in quantum terms by the overlapping of one person's wave function with that of another. The quality and dynamics of that relationship, however, depend on the many variables that can affect any wave system.

Two people who are in the same state, for instance, will have a more harmonious intimate relationship than two people who are in different states, as the wave fronts of their personalities meet in a superposition, one on top of the other or one entangled with the other more or less harmoniously (Figure 9.5). An analogy with musical harmonies (themselves patterns of sound waves) brings this out.

If two musical notes played simultaneously are exactly the same, they can be said to be in the same state, and the result is a single, unified sound. This is equivalent to a harmonious relationship of projective identification, where two people *are* as one.

If the two notes played are an octave apart, their combined sound is harmonious, but it is clear that it is a sound arising from two notes. Similarly, two notes a fifth apart, such as C and G, will produce a harmony, but as the combinations are changed we progress to something like Schoenberg and then further on to simple noise. In the same way, the quality of a relationship depends upon the "ground state" of the persons involved in it.

D. H. Lawrence and his wife, for instance, were said to be inseparable, but their intimacy was sometimes a living hell for both, whereas Robert and Elizabeth Browning complemented each other in nearly every way.

Similarly, people involved in an intimate relationship can share each

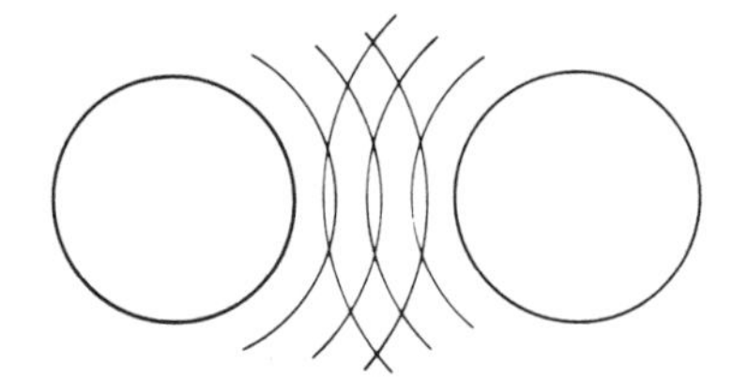

A harmonious relationship—the two people are in the same state.

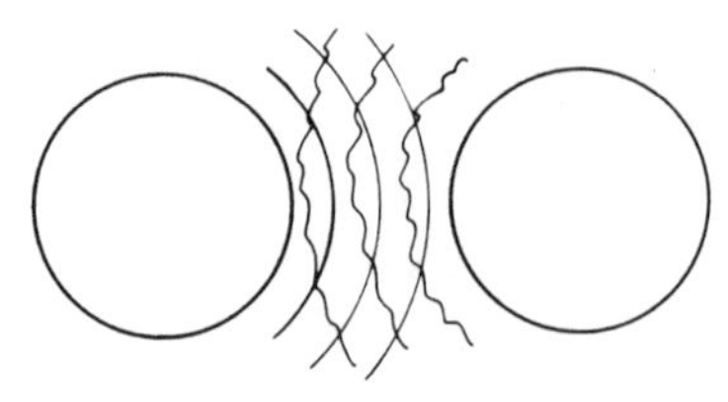

An inharmonious relationship—the two people's wave patterns issue from different states.

Figure 9.5

other's characteristics, as in projective identification, or they can trade characteristics, as happens in role reversals. The latter can be accounted for by a quantum resonance phenomenon in which two coupled quantum systems (or two nonlocally related quantum systems), each with its own characteristic oscillation, suddenly swap oscillations. In this case, I would become you and you me. We can see this familiar pattern exemplified in a situation where one person who has been angry overcomes his anger just at the point where another becomes infected with it, or one member of a couple develops a bad habit just as the other has rid himself of it.

Role reversal through quantum resonance phenomena gives a physical basis for some of the experiences described in catastrophe theory. On a larger and more dramatic scale, this may account for some of the extraordinary "possessions" experienced by psychotherapists who find themselves involved in fantasies once held by their patients—such as the therapist who found himself "traveling" to alien planets and conversing with the natives just at the point where his patient began to realize that such intergalactic activity had been a delusion.[12]

There are countless ways in which quantum wave phenomena both illustrate and explain the dynamics of close interpersonal relationships, mirroring exactly the dynamics of the composite individual self, and it is tempting to declare boldly that there is no real difference.

In so many ways, my relationships with others seem but an extension of my relationships with the subselves of my own self, suggesting that any permanent dividing line between myself and others, between the I and the not-I, is not a very meaningful one. There is no clear way to say where "I" end and "you" begin. In the language of quantum physics: "It must be concluded that macroscopic systems are always correlated in their microscopic states."[13]

Large-scale, individual quantum persons are correlated with ("in tune" with) other quantum persons at the level of excitations on the quantum substrate of their consciousness (on their Bose-Einstein condensates). But they nevertheless remain individuals in their own right. As quantum physicist H. D. Zeh says of more familiar quantum systems, "They still have [some] uncorrelated macroscopic properties."[14] This ability to combine real individuality with definitive relationship is one unique and important result of looking at persons quantum mechanically. Neither individuality nor relationship is lost. Neither is more primary.

Similarly, these same dynamics appear to be further mirrored in the shared conscious experience of people in some group situations—the emotional contagion of a football crowd or a political meeting, or the "group mind" phenomenon noted by psychotherapists treating whole families or networks of close associates. In these, the spoken words of one member of the group seem to express the unspoken thoughts of the entire group.[15]

Viewing the self and its relationships quantum mechanically, we see a spectrum of relationships and communication between the self and others ranging from the private dialogue (overlapping wave functions) between subselves of my own self, through the intimate bond between "I" and "you" with all its possible variations and flavors, and on into the bond that gives some groups the sense of being "at one mind" or "of one heart."

Drawing upon the concept of the wave/particle duality, the notion of a quantum self that is both a self in its own right and a self-for-others cuts a new path through the more familiar dichotomy between seeing the self as *all* or seeing the self as *nothing*.

Seen in its "particle aspect," the quantum self has an important individual integrity. Yet, through its "wave aspect," it is able to be involved simultaneously with other selves and with the culture at large. This lays the basis for both personal identity and personal responsibility, and at the same time for intimacy and group identity. It also suggests a new way of looking at the whole question of personal survival after death.

CHAPTER 10

THE SURVIVAL OF THE SELF: QUANTUM IMMORTALITY

> We must be still and still moving
> Into another intensity
> For a further union, a deeper communion
> Through the dark cold and the empty desolation,
> The wave cry, the wind cry, the vast waters
> Of the petrel and the porpoise. In my end is my
> beginning.
>
> —T. S. Eliot
> "East Coker," *Four Quartets*

During the pregnancy with my first child, and for some months after her birth, I experienced what for me was a strange new way of being. In many ways I lost the sense of myself as an individual, while at the same time gaining a sense of myself as part of some larger and ongoing process.

At first the boundaries of my body extended inwards to embrace and become one with the new life growing inside me. I felt complete and self-contained, a microcosm within which *all* life was enfolded. Later, the boundaries extended outwards to include the baby's own infant form. My body and my self existed to be a source of life and nurture; my rhythms were those of another; my senses became one with hers, and through her, with those of others around me.

During all those months, "I" seemed a very vague thing, something on which I could not focus or get a grip, and yet I experienced myself as extending in all directions, backwards into "before time" and forwards into "all time," inwards towards all possibility and outwards towards all existence.

I joked at the time that I had lost my "particlehood" and my husband told me that I was experiencing projective identification with the baby. Freud would have called it an "oceanic feeling." Whatever, it was both unsettling and exhilarating and through it I lost my lifelong terror of death. It was also, as I said in the Preface, the source of inspiration for this book.

So what have pregnancy and early motherhood to do with death and immortality, or either of these with quantum physics? A deeply felt intuition that they were intimately connected moved me to include a chapter on "Life After Death" in the early synopsis of this book, but for some time after, as the early chapters mounted up and the note cards for that one remained blank, this plan came to seem something of an embarrassment.

Nothing in my own previous way of thinking about immortality, nor anything very definite that I found in the writings of others, seemed to follow from the discussion about the physics of consciousness. Indeed, all talk of consciousness arising from within a quantum process in the brain seemed to outlaw any continuation of consciousness without that brain. And yet, as the outlines of the quantum self, its identity and its relationships, began to emerge from the last two chapters, a wholly new way of thinking about the survival of that self also began to emerge.

At the subatomic level of elementary particles, there is no death in the sense of permanent loss. The quantum vacuum, which is the underlying reality of all that is, exists eternally.* Speaking poetically, we might describe the vacuum as the "well of being." Within this well all basic properties—mass/energy, charge, spin, and so on—are conserved; nothing is ever lost.

Individual particles arise out of the vacuum, exist for a brief while until they collide with other particles, and then either become something new or return to the source from which they sprang. But their brief passage is not in vain. If two elementary particles meet and coalesce, each ceases to exist as itself, but the new particle they become will have the sum of their masses. If a neutron breaks down, its mass, charge, and spin are all conserved in the electron, proton, and antineu-

*Or, at least since the Big Bang and until the final crunch, if there is to be one, and the laws of physics don't require that the vacuum cease even if the universe does collapse into a black hole.

trino that result. Each quantum event that happens leaves its trace, its "footprints on the sands of time."

Similarly, on a larger scale, where the continuity of a pattern or a whole—a group, an institution, a nation—is the object of our concern, the transience of individual components of that whole passes unnoticed, or is at least somehow beside the point.

The individual cells in my body are dying by the thousands every day, but they are replaced by others and my body goes on much as before. The various children now in my son's class at school will grow older and move on, but the school will continue to have a kindergarten, just as there will be daffodils in the garden again next spring when this year's lot have returned to the soil from which they sprang. On a larger scale still, there will always be an England, even if her population changes completely and her cities rise or fall; if not an England, at least nations; if not nations, then a planet; if not this planet, then at least other large bodies that orbit about stars. From *some* point of view, some patterns will always be permanent.

But we human beings, in the course of our daily lives and from the point of view of our private histories, don't see ourselves on the scale of the very small or the very large. We derive only moderate consolation from knowing that our family or our school or our nation or our star will carry on in much the same way after we are gone. That going, the inescapable necessity of our finitude, exists to haunt us every day of our lives. To many it is a shadow that falls over everything they do, to others a scandal that cancels all meaning and value. To escape from this shadow, to deny the scandal, to transcend the finitude itself, most of us either believe in or at least hope for some sort of personal immortality, some survival of ourselves as experiencing, thinking beings. But can there be grounds for such hope?

Traditionally, any expectation of immortality has relied upon belief in the existence of a personal, immortal soul that is independent of the body and survives its death, or on there being some sort of bodily resurrection on the distant horizon, presumably owing to the agency of a transcendent God. A third notion, beloved of spiritualists even today, is that of the "shadow man" or "astral body," a somewhat ghostly and ethereal misty something that detaches itself from the physical body at the moment of death but maintains enough of a shape to be recognizably itself. Each belief, to a greater or lesser extent, flies in the face of modern scientific sensibility.

The idea of an astral body floating free of my immobile corpse, wending its independent way to wherever it is that astral bodies go, is attractive and interesting, and in many ways the most concrete of the three types of speculation about immortality. One can't help wondering, however, about the physics of it, or asking why, despite the best efforts of modern psychical research, no one has managed to detect any "astral influences" hovering near dead bodies.

Equally, the notion that my immortal soul might be freer and happier in the hereafter without its bodily constraint, or that it might one day be reunited with my reconstituted dead body, seems too farfetched for most people to believe with any reasoned understanding. Indeed, the whole notion of an immortal soul, embodied or otherwise, rests firmly upon Platonic and Cartesian dualism, the notion that the soul and the body—consciousness and the brain—are only accidentally connected. But as we have seen already, both split-brain research and the physics of consciousness argue against the separation of mind and body.

So where do we go from here? If, as philosophers and theologians argue, some form of Platonic dualism is a prerequisite for any tenable doctrine of immortality or survival of the self,[1] and such dualism itself is no longer tenable, must we therefore surrender all hope of some sort of meaningful conscious life after death?

The French Christian existentialist Gabriel Marcel, for one, thought not, though he rejected all forms of dualism. "It seems to me," he wrote in one of his essays on immortality, "that we should begin by observing that there can be no question of treating the absolute cessation of consciousness as a fact,"[2] nor the absolute cessation of a loved one as a possibility.

Marcel thought only in terms of ongoing relationships with the dead, even dialogues with the dead,[3] both made possible as a consequence of intimacy between the lover and his beloved while both were alive. While a person we relate to is alive, he argued, we get so inside him that we know what he *would* have said and *would* have thought in given circumstances, and thus we can relate to him as a living presence *now*, not just as a memory.

A position like Marcel's can bring some comfort to the living, offering us an image of how, by refusing to let go, we can keep the dead always with us. But without saying how there can be some physical basis for such a link, some actual mechanics for anchoring the dead in this world, the notion of "creative fidelity," as Marcel calls our loyalty to the dead, can be little more than wishful thinking, a salve for the pain of loss. There is no way to see how it can actually keep the dead alive in any sense useful to the dead themselves. There are no grounds for the dead having continuing *experience.*

In a classical view of the self, a view that sees the self as an isolated individual essentially cut off from others and arising solely from neural pathways within its own brain, there is no way beyond this impasse. In classical terms, there can be no "physics of immortality" based upon intimate relationship because there is no physics of intimacy. But for a quantum self, things are very different.

Intimate relationship, the type of relationship that gets inside the self, that influences and even defines its being from within, is the sine qua non of the quantum self. Viewed quantum mechanically, I am my relationships—my relationships to the subselves within my own self and my relationships to others, my living relationships to my own past through quantum memory and to my future through my possibilities. Without relationship I am nothing.

In considering whether there is a quantum view of immortality that meaningfully binds the dead to this world through relationship—Marcel's "creative fidelity"—we are asking whether there is any physical basis for integrating another's past into our "now" in such a way that that other, though "dead," is really here with us now, laughing as we laugh, planning as we plan, loving as we love.

We are really asking whether another's past life can be reincarnated through his bond with us. Is this any different from asking about the bond between my own past and my living present?

Through the process of quantum memory, where the wave patterns created by past experiences merge in the brain's quantum system with wave patterns created by present experience, my past is always with me. It exists not as a "memory," a finished and closed fact that I can recall, but as a living presence that partially defines what I am now. The wave patterns of the past are taken up and woven into now, relived afresh at each moment as something that has been but also as something that is now being.

Through quantum memory, the past is alive, open, and in dialogue with the present. As in any true dialogue, this means that not only does the past influence the present but also that the present impinges on the past, giving it new life and new meaning, at times transforming it utterly. A personal example may help to make this more concrete.

As a baby and young child I was separated from my mother a great deal, often for months at a time. For three years I didn't live with her at all and saw her only on occasional, fleeting weekend visits. I missed her dreadfully, cried often for her, suffered very early depression about her absences, and often plotted ways by which I could escape from my grandparents' home and get back to her.

These early separations marked my childhood, without question, but they also cast a shadow over my adult life. The child within me (my child subself) was woven into the patterns of relationship I experienced as a teenager and adult.

For years I suffered dreadful insecurity in my relationships with others, wondering whether they really wanted me, whether they would reject me. If anyone did seem to love and want me, I experienced what the psychologists call separation anxiety whenever he or she was out of my sight. I couldn't easily tolerate the freedom of those I loved, and this in turn made me seem a suffocating presence for them—often inviting the very rejection I feared.

With my rational mind, I could at times see the pattern of clinging suffocation followed inevitably by rejection that blighted my adult relationships, but no amount of rational insight seemed to alter it. The rejected child within me was there as a living companion in every liaison I made. It was only when my own children were born that everything changed.

During the first night in the hospital after my daughter was born, I missed my mother with a greater and more painful urgency than I had ever experienced before. I wanted her to be there with me and my new baby. As so had happened often before, she was not. But then something began to happen. Even during that first night I felt myself becoming that mother—not simply the mother of the new baby I held in my arms, but also the mother of the baby within myself. As I held my own baby close, I also embraced the baby within me. I loved her and reassured her that I would always be there.

As the months of my daughter's infancy passed, I realized repeatedly that in being a good mother to her I was also being a good mother to myself. When she cried in the night and I went to her, I felt my own baby self crying out and she, too, was comforted. There were no more lonely nights for that baby within, no more painful separations. Her unhappy infancy was taken up into the present, interwoven with all the ministrations lavished on my own daughter, and she became secure.

Reincarnated through quantum memory, the baby within received a new start in life. She was "born again."

Many parents have experienced some form of identification with their young children through which they have reached stunted parts of themselves. This is one reason why the experience of parenthood so often leads to growth and greater maturity. Similarly, as mentioned earlier, genuine reliving of past events through quantum memory is basic to the growth mechanisms unleashed by successful psychotherapy. But in considering the problem of immortality, the question is whether we can similarly embody, and hence reincarnate, the past of another. Can someone I love be born again through me in the same way that aspects of my own past self are reborn?

On a daily basis, while we are both alive, the answer seems an obvious yes. Through intimacy, the lover and his beloved (or the mother and her child, the members of any close association or group) are so entwined, their wave functions so overlapped, that neither can tell "who was that I whose mine was you." Each is, to some extent, the stuff of which the other is made. Since each is made, in part, by those elements of his past that are woven into his present, each of us carries within himself both his own past and aspects of the pasts of those with whom he is intimate.

Thus, just as I am in constant dialogue with my own past, so I am in constant dialogue with some of my husband's past—those elements of it that he has brought into our relationship. In relating to him I am relating also to aspects of his babyhood, to his parents, and to his childhood in Canada. In being a good mother to the child within myself I am also a good mother to his child within, that child which has been a lived part of our adult relationship.

In a quantum view of the self, there can be no hard-and-fast distinction between my own past and that of another with whom I am intimate. Indeed, through me, through relating to me, that other might achieve some dialogue with his own past that he might otherwise not have had. Thus through me, at each successive moment of my life, elements of the other's past are reincarnated, just as my own past is reincarnated moment by moment—reincarnated in my present and present ever after as part of the tapestry of my being.

The others whose past selves are woven into my own self may be people with whom I now share an ongoing intimate relationship, but they may also be people who have lived before me—parents, grandparents, heroes, and historical figures, each of whom has in some way impinged on (overlapped with) my consciousness or gone to make up the consciousness of someone who has. I am, in part, my parents and my grandparents, and through them generations of progenitors whom I personally have never known.

Equally, through folk memory and the way that it, too, is taken up into my quantum memory and interwoven with my present, I can be, if I am an American, in part George Washington or Abraham Lincoln or Jack Kennedy. That is, Washington's honesty, Lincoln's sense of fairness, and Kennedy's youthful enthusiasm are all part of the warp and woof of my own being to the extent that I have valued and revered (formed a kind of intimate bond with) these qualities in them. This is the physical basis of our being historical beings. We are literally interwoven with history, and at the same time, history is in dialogue with us.

As Marcel says about his link with the pasts of others, "I must think of myself not merely as somebody thrust into the world at a moment of time that can be historically located, but also as bound to those who have gone before me in some fashion that cannot be brought down to a mere linkage of cause and effect."[4]

Understood in terms of quantum memory, this bond with those who have gone before—with the dead—like the bond with the pasts of those living people whom I love, is not a bond of mere "memory." It is not that I *recall* them, but that I *am* (in part) them. Through me, through the fact that aspects of their being are woven into my own, they are reincarnated—taken up into my life to live as I live.

But surely, we want to argue, the dead *can't* "live as I live." Perhaps it is possible through quantum memory that a dead person's past life is now part and parcel of my own living life, but what distinguishes my life as living is that I have ongoing experience. I am aware of myself as alive, and I have a future. I marvel at the beauty of the morning sunlight reflecting on the canal beneath my window and I will most likely do so again tomorrow. Surely the dead have no such experience, and certainly no future.

But such arguments only expose the tenacity of our old, prequantum way of looking at the self—not just our way of looking at the survival of the self, but also our way of looking at the living self as it exists through time and relates to others.

In a classical view of the self—a view that, if not dualist (my mind and my body are separate entities), is necessarily materialist (I am my brain)—there is no way to account for the persistence of the self through time, never mind after death, and no way to account for intimate relationship.* By contrast, in a quantum view, there is no way to draw any sharp distinction between my persistence through time, my close relationship to others, and my survival after death. Neither isolation nor death has any clear-cut meaning.

In a quantum view, my relationship to myself across time—the wedding of my many accumulating subselves through quantum memory—is very like my intimate relationship with another at any moment. In both cases "I," as I am "now," arise from a tapestry of interwoven patterns (oscillations) on the quantum system in my brain. Some of these patterns emerge from neural pathways laid down in my own past, and some emerge from nonlocal correlations with patterns on the quantum system in the other's brain,† but both are woven into "me."

I am I (the union of all my subselves), but I am also I-and-you (the union with you). If I die, it is true that there will be no more ongoing dialogue within myself—within that inimitable pattern that arises from the combination of all my past, all my awareness and experiences, all my relationships, all my genetic material, all my bodily idiosyncrasies. In the language of quantum physics, I will have no more "particle aspect." But the part of myself which I have brought into relationship with you, my "wave aspect," the I-and-you, will continue as part of your dialogue with yourself and others.

Thus as long as you have experience, I-and-you have experience; as long as you have a future, I-and-you have a future. While I am alive, fully alive in the commonsense understanding of that word, I create myself; after I am "dead," when I have ceased to be a "particle," you create me.

In principle, there is only a shade of difference in the way I create myself (weave myself) across time while I am alive and the way in which I continue being created by you after I am dead. After all, if you and I are in an intimate relationship, I am being partially created by you

---

*See the argument in Chapters 8 and 9.

†The notion that the overlapping of quantum systems arising from within two separate brains is due to nonlocal correlation effects is "grounded" speculation. Such correlation effects do exist between photons in separate laser beams, and the quantum vacuum is full of them. But there may be some other explanation for the overlap of persons, based on physical processes that we don't yet know.

at every moment as I live. There is no sharp dividing line between these stages of my personal evolution (the period of my life time and the period after my death); they are both aspects of one ongoing process.

While I am alive, I am changing and growing from one moment to the next. This is true of both my body and my character, indeed of the whole pattern that is "me."

Each day thousands of neurons in my brain and tens of thousands of cells in my body die, to be replaced by others of their kind,* but "I" continue. In the same way, the "I" that exists now, though woven in part from the cloth that was "me" yesterday, is an evolved person in whom yesterday's "I" is reincarnated. My childhood self no longer exists exactly as the child that I was, and yet it lives on in me, partly to make me what I am and partly to experience its own growth through me. (I am the mother of my own child within.)

Thus I am always investing my future in another. While I am alive, that other is "me"—my many selves that I am becoming; after I am dead, that other is you. But my growing does not cease. The process of my becoming continues.

Insight into the process of becoming and the continuity of persons through the processes made possible by quantum memory is one of the deep and abiding visions that quantum physics holds out for our way of being in the world. It touches to the core both our sense of ourselves as persons through time, and our understanding of our relationships to ourselves and to others—both within time and beyond time. It places us in the world, not just here and now, but always.

Like an electron, each of us is a "point source" in space and time (our particle aspect) and at the same time a complex pattern woven from our comingling with others (our wave aspect). We, too, are patterns of active energy, patterns arising from within ourselves (our genetic codes, the structures of our bodies, our senses, and all our experiences) and from beyond ourselves (the structures and experiences of others, many of whom have lived before us and others who will live after). For each of us, there is no clear way to say where that pattern begins or ends. "In my beginning is my end," but also "in my end is my beginning."[5]

---

*This is less true in the case of brain neurons. Once the brain reaches its mature size, there is little replacement of dying neurons.

The idea that all life is an ongoing process of which we are a part is not in itself new. Anyone who is close to Nature and her processes can see that it is obviously the case. No new physics of the person or of immortality is required to see that my body is made out of atoms that once were stardust and that one day they will find their home again among distant galaxies. I am made of the stuff of which the universe is made, and the universe shall be made of me. It is equally clear that I received half my genetic material from my mother and that a fair proportion of that has been passed on to my own daughter.

But with a quantum view of the self, and an understanding of the way my own self is literally (physically) woven into the selves of others (has become a pattern on the quantum substrates of their consciousness), my place in this process becomes more personal and more abiding. I am not just a link in the chain of process, a bridge that others cross on the road to the future—these are Newtonian images taken from the notion that time is a series of successive moments.

Rather, with a quantum view of process, it becomes clear in a new way that "I," not just my atoms or my genes, but my personal being—the pattern that is me—will be part and parcel of all that is to come, just as it is part of the nexus of now and, indeed, was in large part foreshadowed in the past.

Just as there is no space or time between two separate laser beams (their wave patterns interfere across space and time),[6] so there is no real division in space or time between selves. We are all individuals, but individuals within a greater unity, a unity that defines each of us in terms of others and gives each of us a stake in eternity.

Understanding this, understanding the full reality of the extent to which we are all physically interwoven, requires a revolution in our whole way of perceiving ourselves and our relation to others. It is a revolution required when we apply quantum concepts to the nature of the self. We know that quantum physics calls upon us to alter our notions of space and time, but now we have to accept that this touches each and every one of us at the core of his personhood.

There is something deeply feminine about seeing the self as part of a quantum process, about feeling in one's whole being that I and you overlap and are interwoven, both now and in the future. Selecting things out, seeing them as separate, naming them, and structuring them logically are male attributes. They follow, if you like, from the "particle aspect" of our intelligence. Seeing the connections between things is more feminine. It mirrors the "wave aspect" of the psyche.

My own insight into the truth of the process view came through the experience of pregnancy and early motherhood, but one needn't be a mother, or even a woman, to appreciate the essential connectedness of quantum theory and what it is telling us about ourselves as quantum systems. We all, both men and women, have a feminine side, a "wave aspect," an aspect that surrenders rather than grasps, that "gives itself up" to things beyond the nuclear self rather than concentrating on building boundaries around that self. This is the side we must cultivate if we are to transcend isolation and the consequent, and needless, terror of death.

But the surrender called for to make the most of quantum relationship and quantum memory is not a passive surrender. It is not the surrender of the mystic or the dropout.

The I-and-you of intimate relationship exists only to the extent that I relate to you in the first place, and all relationship is an effort. Two chemical substances often won't give up their inert existence to combine unless they are excited by the addition of heat. They must overcome the threshold of their potential energy. In the same way, just as I get my own self "together" only to the extent that I work on myself, to the extent that I strive energetically and with concentration to integrate the many subselves within myself, so I am interwoven with you only to the extent that I commit myself to that task.

I must focus all my passion, my loyalty, and my care onto the ongoing, evolving process of I-and-you—both the I-and-you of more personal one-to-one relationships and the I-and-you in its broader sense. By the latter I mean the family, the group, the nation, life as a whole—each of the many layers of relationship where my own being can comingle, overlap, and entwine with that of others. To the extent that I do that, I ensure my place in the ongoingness of things.

The kind of surrender required to make the most of quantum process is like Christ's saying, ". . . and whosoever will lose his life for my sake shall find it." On a quantum view, he who would find himself a place in eternity must fully wed himself to life's processes of relationship now.

I am reminded of the old song that tells us that we can't get to heaven in an old Ford car because an old Ford car won't get us very far. Equally, we can't secure much of a place in the future lives of others without a fair degree of commitment and responsibility towards relationship now. We only get out what we put in. We survive only to the extent that we have lived.

> We must be still and still moving
> Into another intensity
> For a further union, a deeper communion. . . .

CHAPTER 11

GETTING BEYOND NARCISSISM: THE FOUNDATIONS OF A NEW QUANTUM PSYCHOLOGY

> Emotionally shallow, fearful of intimacy, primed with pseudo-self-insight, indulging in sexual promiscuity, dreading old age and death, the new narcissist has lost interest in the future.
>
> —Christopher Lasch
> *The Culture of Narcissism*

A fear of intimacy, dread of old age and death, and loss of interest in the future seem poles apart from the sorts of relationships and concerns bound up with the survival of the self as understood in quantum terms. And yet, such disturbing anxieties are a familiar part of our emotional landscape, and a need to transcend them one of the most urgent challenges that we face, both as individuals and as a culture.

We in the West, in the twentieth century, live largely in what can be described as an I-centered, or now-centered, culture. It is what Christopher Lasch and others have called a narcissistic culture. It is a culture that stresses the importance of "I" and "mine." The individual, his experiences, his feelings, his "happiness" are the focus of attention, truth, and value.

If something makes me feel good, it must be a good thing. If something is true for me, it must have a kind of validity. "Every truth is a truth for someone," and my point of view has special status as my window on reality. My experiences are what truly count and I should have as many of them as I want. I must be "true to myself." I must "get to know myself" and "do my own thing." The whole ethic of all this self-centered self-importance was summed up in the "Gestalt Prayer" that was so central to the 1960s self-awareness movement, though its appeal was by no means limited to the followers of Gestalt psychotherapy nor to the decade of the sixties.

> I do my thing, and you do your thing.
> I am not in this world to live up to your expectations
> And you are not in this world to live up to mine.
> You are you and I am I,
> And if, by chance, we find each other, it's beautiful.
> If not, it can't be helped.[1]

The consequent selfishness, shallowness, alienation, and downright unhappiness of all these cosseted individuals are familiar concerns to many of us in our daily lives. The sad irony here is that in a culture that lays so much stress on the individual, his own sense of personal worth and power is so diminished. As many psychologists have noted, narcissism is more about self-hatred than self-love,[2] and is commonly associated with feelings of emptiness, worthlessness, personal disintegration, and pent-up rage (Table 11.1). These symptoms are the source of a great deal of social tension and personal pain, and have generated a whole popular literature that includes books like Lasch's *The Culture of Narcissism* and, more recently, Allan Bloom's *The Closing of the American Mind.* Both describe in graphic detail the side effects of too much emphasis on what I would call the particle side of our being.

Narcissism is a disease of relationship, a disease that springs from a failure to make meaningful relationships to oneself and others. Its opposite is an attitude towards life that stresses the importance of commitment, involvement, love, sacrifice, and even, at the greatest extreme, perhaps martyrdom. It is an attitude that takes the individual beyond himself, beyond his own isolated islands of experience, his own feelings and reflections, and grounds him in the wider context of life and relationship.

Such an attitude has existed in past, more religious times, but it is not a dominant theme in our own culture.

Obviously, not all people in Western countries are leading empty and narcissistic lives. Many have fulfilling relationships and know the meaning of commitment, intimacy, and sacrifice. A great many more have such things as an ideal. But our *model* of ourselves, the psychologi-

TABLE 11.1
THREE EXPRESSIONS OF NARCISSISM

False Self – The Defences the Narcissist Develops to Make Himself Feel Better	*Symptomatic Self – Feelings to Which These Qualities Give Rise*
Reliance on achievement	Vulnerable to shame, humiliation
Perfectionism	Hypochondriacal, psychosomatic
Grandiosity-omnipotence	Worthlessness, self-depreciation
Pride	Isolation, loneliness
Entitlement	Depression, inertia, work inhibition
Self-involvement	
Manipulation and objectification of others	

Real Self – The Actual Qualities of the Narcissistic Self
Feelings of emptiness, void, panic with enfeeblement and fragmentation of the self
Archaic demands of rapprochement: Merger, twinship, mirror and idealization transferences
Feelings of rage and hurt at the empathic failures of the archaic demands
Searching for, discovery and development of the real self: Innate capacities, identification, ambitions, and ideals

Table adapted from Johnson, Stephen, *Humanizing the Narcissistic Style*

cal mirror into which we look when we want to know who we are and how it is valuable to behave, is a narcissistic model that follows necessarily from our existing psychology of the person. If we want to grow beyond that model, we must grow beyond the psychology on which it is based.

Our present psychology of the person rests almost entirely on a model of the self as a thing that exists in isolation. Though it has many sources in the different strands of our post-seventeenth-century Western intellectual tradition, most especially the decline of traditional religion and the rise of modern science—the philosophy of Descartes and the physics of Newton—this model was really focused as a coherent and consistent psychology of the person by Freud. It is through a vague familiarity with his work that so many people have been affected by it. This influence is so great that it would be impossible to separate our current understanding of ourselves from the wider framework of his early vision.

And the conceptual core of Freud's vision was that the world consists of selves and objects, each self a stranger to all others because of an essential separateness.

As a leading British Freudian psychoanalyst expressed it at a conference I attended recently, "I am a self to myself, but an object to others. To others I am a thing, a 'what,' and others are objects to me."[3] The whole of Freudian psychology is a psychology of the individual and his "object relations."

Moreover, Freud's early stress on the sexual origin of all neurosis and the dominance of the pleasure principle portrayed human beings as selfish creatures bound by instinct and an urge to feel good, while his insistence on the passive role of the analyst reinforced the patient's isolation from testing, and potentially nourishing, relationships. Such features in Freud's thinking produced a reaction among his followers, resulting, at last count, in some 250 different attempts to take his early vision beyond its narrow confines.[4]

Some of Freud's successors, like Alfred Adler, tried to stress the social nature of human beings and to encourage more responsible and committed community attitudes. Others, like Carl Rogers, concentrated on the importance of a two-way relationship between therapist and patient as an experience of mutual growth. Group therapists emphasized the importance of whole networks of relationship; humanistic psychologists stressed the importance of immediate experience—insight, ecstasy, communication, often induced through drugs or meditation or touching of one sort or another; and the existential psychoanalysts sought to develop "authenticity" and stressed our being-in-the-world. But all, regardless of their aims, stoked the fires of narcissistic overattention to self.

"All psychotherapies," says American psychiatrist Jerome Frank, "despite their diversity, share a value system which accords primacy to individual self-fulfillment or self-actualization. The individual is seen as the center of his moral universe, and concern for others is believed to follow from his own self-realization. . . . Our psychotherapeutic literature has contained precious little on the redemptive power of suffering, acceptance of one's lot in life, filial piety, adherence to tradition, self-restraint and moderation."[5]

The isolation of the self by psychoanalysis and psychotherapy was further underpinned by the growth of medical psychiatry as a scientific discipline on an equal footing with brain surgery or general medicine. Psychiatrists treat the person as a separate physiological system, and regard anything psychiatrically wrong with him as issuing from imbalances within that system—chemical imbalances in the brain, which can be corrected with drugs.

Jung's work—his emphasis on the collective unconscious, his notion of synchronistic connections between people and events, his wider definition of the self to include shared archetypes and images of unity, totality, and immortality—is in many ways a glaring exception to these trends in psychoanalysis and medical psychiatry. Strangely enough, though, his more transpersonal psychology has had curiously little impact on the central ethic of psychotherapy.[6]

And that ethic, the general ethos of I-centeredness, has left its mark on the thinking of people with no direct experience of psychotherapists and their ilk as much as Newton's physics has colored the thinking and self-images of people with little or no direct experience of scientific laboratories. These things are in the air; they are the stick against which we measure ourselves and our behavior. They have become the basis of our "folk psychology."

But if the goal of psychoanalysis and psychotherapy was the self-fulfillment of the individual, their failure has been their inability to bring about any such thing. People are not, on the whole, more self-realized or self-fulfilled than they were when Freud began his work.

If anything, loneliness and alienation—alienation both from self and from others—are problems of our time more than of Freud's, as is the narcissism that underpins much of them. Several analysts have noted that a far higher proportion of the problems that cause patients to seek their help today have their roots in narcissistic personality disorders. So far as psychology's role in this is concerned, as Allan Bloom has pointed out, "The only mistake was to encourage the belief that by becoming more 'inner directed,' going further down the path of the isolated self, people will be less lonely."[7]

The self thrown back entirely on itself, with nothing but itself as a source of meaning, truth, and value, has no nourishment on which to draw. It is like a plant that has been potted in a box in the garden shed rather than outside in the soil and sunlight. Very soon its roots dry up and its leaves wither. In Bloom's words, "There must be an outside for the inside to have meaning."[8] There must be something beyond ourselves to give those selves a sense of what we are about.

There is much to be said for the details of Freud's work and its development by his followers. His groundwork on the interpretation of dreams, his articulation of important defense mechanisms (suppression, rationalization, projection, and so on), and his basic analysis of the stages of development all have valid and lasting application to our understanding of the dynamics of the individual psyche.

Equally, there is obvious truth in medical psychiatry's claim that some diseases of the self are the result of brain tumors or chemical imbalance. But on their own, such insights are without a meaningful context. They are not adequate as a *paradigm* of how human beings function.

I think that an understanding of the quantum nature of the person, solidly based on the quantum mechanical nature of consciousness itself, can give us such a paradigm and thus lay the foundation for a wholly different, nonnarcissistic psychology of the person. Some of Jung's followers have recently argued that a wider appreciation of quantum reality would give his own broader vision of the self a more scientific foundation and thus a more general acceptance.[9]

The essential interwovenness of quantum reality, including ourselves as quantum persons and the quantum vision that our place in the here and now, never mind in eternity, depends upon our deepening relationships to others and the commitment required to achieve them. This requires a radical turnaround from our accustomed egocentric, and hence alienated, way of looking at things. The concept that I *am* my relationships requires a similar turnaround. A closer look at the nature of commitment itself, the driving force that binds us to any relationship, can bring the implications of this new "quantum psychology" into sharper perspective.

The essential basis of any commitment is that we are defined by certain things, that they are in some sense that of which we are made. A sense of commitment requires an intimate sense of "at-homeness" with those things to which one is going to be committed—whether these are spiritual values like truth and beauty, interpersonal or social relations (friends, family, community, country), or Nature herself. An uncommitted person says, typically, of such things, "It has nothing to do with me." A narcissistic person *feels,* "It has nothing to do with me."

There is no place for commitment in Freudian psychology, just as there is no conceptual framework for interpersonal relationships. The word *commitment* does not appear in Charles Rycroft's *Critical Dictionary of Psychoanalysis.* [10] In its place is Freud's notion of "cathexis," an attachment of the libido to some object, whether internal or external.

The cathecting self finds itself obsessed by something, a quantity of its energy directed towards that object in much the same way as an electric charge or a magnetic force is directed towards an opposite polarity. It's a mechanistic image and, like so many of Freud's images, self-centered. It's all about the internal, shifting balance of the isolated psyche's own energy reserves as the psyche reflects objects back on itself. As Rycroft reminds us, "Psychoanalysis . . . discusses objects and relationships only from the point of view of a single subject."[11]

Of course Freudians do discuss the importance of interpersonal relationships, of commitment, mediation, conciliation, and respect for other persons, but in this, their own experience as human beings contradicts their theory and exposes its weakness. How do we mediate or conciliate with objects? How do we respect them, and what basis would we have for any commitment to them? They are entirely *other* than ourselves.

Similarly, Freud's model of the person lays no basis for a commitment to Nature or to spiritual values. His "scientific psychology" aims at an understanding of the self as a biological entity akin to plants and animals, but his mechanistic interpretation of biology itself gives a deterministic and somewhat brutish impression of both ourselves and our biological fellows.

Animals, he argued, including ourselves, are driven in their behavior by the inseparable instincts for sex and aggression. In humans, these instincts control the dark, hydraulic forces of the id and are the underlying, unconscious causes of all that we do. They bind us to Nature and trap us there, beasts among the beasts.

For Freud himself, there can be no question of committing ourselves to Nature, to the beast within. The task of consciousness—the ego—is to suppress and transcend these dark instincts with the power of rationality. Hence his famous dictum: "Where Id was, there Ego shall be." Yet this very renunciation, on which our civilization rests, involves us in a tragic and impossible conflict.

The spiritual values—such as love, truth, beauty, and inquiry—that underpin our lives and raise us above the beasts arise, for Freud, out of the sublimation (transformation) of our more primitive, natural instincts. The impetus for this sublimation comes from the dictates of the superego—the unconscious internalization of society's values and behavior patterns as learned through our parents. These dictates are not our own; they are not part of our basic nature. Rather, they are imposed from outside precisely in order to curb that nature. They make us feel guilty and set us at war against ourselves.

"If," says Freud, "civilization imposes such great sacrifices not only on man's sexuality but on his aggressiveness, we can understand better why it is hard for him to be happy in that civilization. In fact, primitive man was better off in knowing no restrictions of instinct. To counterbalance this, his prospects of enjoying this happiness for any length of time were very slender. Civilized man has exchanged a portion of his possibilities for happiness for a portion of security."[12]

Thus our spiritual values are a prudent, though expedient, compromise. We can feel no basic commitment to them. They are not the stuff of which we are made, but rather the (considerably uncomfortable) clothes with which we cover our true nature. Taking them off releases the beast in us and destroys our civilization; wearing them stifles and distorts us.

The existentialist emphasis on commitment, which many psychoanalysts credit with broadening the scope and shifting the emphasis of their work, suffers many of the same defects, though they are expressed differently.

For both Sartre and the early Heidegger *(Being and Time),* the unbridgeable distance between the self and others gives interpersonal commitment an arbitrary and self-centered quality. I am committed because *I* choose to be, not because the other, who is just a mirror that narcissistically reflects my existence back to me, in any way solicits that commitment. It is *I,* through my choosing, who lend meaning and value to the commitment, and *I* who benefit by thus exercising my freedom.

The nature of the existentialists' "I" lends a further whimsical and arbitrary quality to their understanding of commitment, or choice, especially as this affects commitment to spiritual values or to Nature.

For Sartre, like Freud, there can be no question of a commitment to the natural in oneself—not because that nature is brutal and selfish in a way that undermines the self's own best interests, but because its very existence is an illusion and an excuse. Existential psychoanalysts want to distance themselves from such notions as human nature, hereditary dispositions, or character—an attitude expressed in the famous "existence precedes essence" of Sartre.[13]

"If existence really does precede essence," he says, "there is no explaining things away by reference to a fixed or given human nature. In other words there is no determinism, man is free, man is freedom. On the other hand, if God does not exist, we find no values or commands to turn to which legitimize our conduct. So, in the bright realm of values, we have no excuse behind us, nor justification before us. We are alone, with no excuses."[14]

"I" am nothing but my choices, my utterly free and utterly necessary capacity to make choices and to create values, but the precise choices themselves are without necessity, or even foundation. There is no *reason* for them, no underlying natural or moral imperative that they be of one sort or another. Thus I might choose commitment to a given person or self-designed set of values today, but equally well I might choose some other tomorrow. I define myself as I go along, and nothing ever has to stick. I have no past.

This denial of the past is a trait in modern culture that in part Sartre simply reflects, and to which in part his brand of existentialism has contributed. It is a denial that underlies the narcissist's loss of interest in the future.

"The narcissist has no interest in the future," argues Christopher Lasch, "because, in part, he has so little interest in the past." Deprived of the past's vast psychological storehouse of experience and memory, he suffers from an "impoverishment of the psyche and also from an inability to ground [his] needs in the experience of satisfaction and contentment."[15] With so little on which to draw, he feels empty and confronts the future with lethargy and depression.

With a quantum view of the self, both the nature of the self and its interpersonal commitments are radically different.

In the first place, the quantum self has, in Sartrian terms, both essence and existence. I do exist as a person with an identity, a character, a style—some of which is influenced by hereditary dispositions—

and the things that I do and the relationships that I make do "stick." This follows from the physics of consciousness and the ongoing dialogue between the brain's quantum system (its Bose-Einstein condensate) and its neural pathways.

Events in consciousness (excitations of the Bose-Einstein condensate, which is the physical basis of consciousness) feed back into the brain's neural system, laying down new pathways or reinforcing old ones. They are literally "etched" on the brain. These pathways in turn can transmit signals back to consciousness at any time as part of the quantum memory system, where their excitation patterns overlap with those of new experience or pockets of past awareness to "weave" my evolving self.

Thus I do make myself as I go along, each new relationship does alter and partially redefine the self that I am, but I am never a blank slate as Sartre claims, nor is my past ever lost. Indeed, that is what gives commitment its meaning and purpose. If commitment is the process by which something becomes a part of oneself, that self must be a thing that is capable of taking things into itself and retaining them there. It must have an "essence."

Equally, commitment understood in quantum terms cannot be a solitary thing, as are both Freudian cathexis and existential choice. It is not something directed *towards* others, nor *projected upon* others, but an act of fidelity bound up *with* others as a fundamental part of the self's own definition, a fundamental part of its very nature as a system always engaged in creative relationship. If I am committed to you, I recognize that you and I are stuff of the same substance, that your being is entangled with mine forever after. Such entanglement has its physical basis in the uniquely quantum phenomena of nonlocality, the distant correlation of apparently separate quantum systems, and coalescence, the ability of boson systems to overlap and share an identity.

Furthermore, the human nature that I possess as a being whose consciousness rests on these quantum phenomena is a nature that I share with all other living things whose body cells contain quantum systems (Fröhlich-style Bose-Einstein condensates). Indeed, ultimately, I share this nature with all other boson systems, which, even at the level of elementary particles, have the need to make relationships as a basic quality of their existence. Bosons themselves are nothing but "particles of relationship."

Thus as a quantum self I have a basis for commitment to the whole world of Nature and material reality. We are *all,* basically, stuff of the same substance. And the same can be said of spiritual values such as love, truth, and beauty.

In a quantum view, these values are not mere projections of myself, sublimations of some dark and unacceptable side of my nature, as Freud would say; nor are they something that I create *ex nihilo,* as Sartre would say. They have a being of their own that arises out of their basic nature as "relational wholes"—things that in their being create relationships—and this nature happens to be the same as my own.

Love, most obviously, draws things (such as art objects or other values) and people together. As I noted about Plato's view in Chapter 7, where there is the lover and his beloved there is also a third thing, which is the love between them. The love itself has a kind of existence, which arises from the relationship.

Similarly, beauty or art is a relationship that draws together previously separate elements into a new whole, which then has a being of its own—e.g., van Gogh's painting of the peasant's shoes, which draws together the peasant, the earth, the sky, the peasant's labor, the history and meaning of all labor, and so on. Truth is the creation of a correspondence, a relationship, between elements of reality and between those elements and reality itself. And as Heidegger rightly says, truth and beauty, or truth and art, cannot be separated from each other, and neither can be separated from the expression of relational holism.[16]

In my own being, which draws its very existence from the creation of relational wholes, I am by nature a creature that is stuff of the same substance as love, truth, and beauty. Not because I create them, but because the nature of my own consciousness is synonymous with the nature of their meaning. Through my own being I have the capacity to act as midwife to their expression in this world, and they in turn mold and make the self that I am.

The same could be said of any spiritual values, all of which share the common quality that they create relationship, and thus are stuff of the same substance as myself. There is a firm basis for commitment to them.

All quantum systems in the universe, including ourselves, are entangled (correlated and interwoven) to some extent. Even the quantum vacuum is full of correlations.[17] Such basic entanglement is the essence of quantum reality. But these same systems also have the potential for *more* entanglement, for more and deeper relationships, and that potential is an important aspect of a psychology based on the quantum nature of the person. It gives it a dynamic.

The basic, weak entanglement of all quantum systems gives us a *basis* for commitment. It is given to us as a birthright. But the potential for more and deeper entanglement, which depends upon the degree of similarity achieved between these systems, gives us a *motivation* for commitment. It spurs us on and gives us a natural direction in life.

Each of my intimate relationships, however brief, does get "inside" me, does add at least some small thread to the tapestry of my being. But just as diverse small threads do little to make a recognizable pattern in a woven tapestry, so many brief intimacies or small forays into involvement do little for the integration of myself or my union with others. So dispersed, I lack a theme, a central core that either I myself or others can recognize. I have little on which to build further relationships.

This is the situation of the narcissistic personality. Unable to feel any basis for a commitment to others, to Nature, or to any coherent set of values, and thus unable to sustain any deep relationships, he experiences both a fragmentation of himself and an isolation from wider communion.

But if I commit myself to others (or to Nature or to some spiritual value), I become more entangled (more at one) with them through a form of repetition. Each day in various large and small ways I renew my relationship with the other, perhaps through added contact and more shared experience, through memory and reflection, or through the influence my commitment has on other aspects of my thought and behavior.

I repeatedly take the other into myself, thus reinforcing excitation patterns on the quantum substrate of my consciousness, and with each repetition his being becomes more a part of my own, more interwoven with other aspects of my being. Our identities overlap and our personal qualities become more correlated. Both the relationship and myself grow accordingly. "I" become an extended self, a much larger part of which is I-and-you.

Equally, this quantum view of commitment sheds new light on the implications of breaking one's commitments. If I do break a commitment, it is not just the other whom I injure, but also myself. A broken commitment is a withdrawal from the defining relationship created by that commitment, and what I lose is literally a part of myself. I lose that part which is the ongoing process of I-and-you; this process ceases to be a unifying thread in my life, a growing point. It becomes, instead, like a forgotten portion of my childhood, a subself that is largely disengaged from the central integrating structure of my self. I become fragmented.

But because nothing is ever entirely lost, because every committed relationship is woven into my being forever after, there is always the potential to renew a broken commitment. I can reestablish a creative dialogue with that past I-and-you which causes it to be reborn, albeit in a different form. Hence the proverb of the prodigal son.

Because the basis of commitment is, at its most primary level, a sense of at-homeness with the other, a sense that he is "someone like me," the ease with which we make a personal commitment is often greatest in cases where there are already many shared attributes—such as commitments to members of our own family with whom we share genetic dispositions and with whom we have a large pool of shared experience, or members of our group or culture with whom we share habits, language, and patterns of thought.

Such incipient similarities make the effects of commitment more immediate because there is already some degree of correlation and overlap among selves that share a history or a tradition. Research has shown, for instance, that the most stable marriages are those between partners with similar personalities and backgrounds. They are already to a large extent stuff of the same substance. This is even more obviously true of mothers and their babies, for whom projective identification (the sharing of an identity) is the norm, and of identical twins, whose lives seem almost eerily correlated at a great many levels.

But except in these extreme cases where considerable overlap and correlation exists almost as a birthright, some active work is required to sustain and deepen relationships even within our group or culture. This might take the subtle form of an adherence to or renewal of certain values cherished by the group or culture—such as admiration of physical or mental achievement, a desire to help others less well off than oneself, a high regard for personal freedom—or it might be expressed through more organized behavior.

The observance of rituals, anniversaries, and public holidays; the repetition of national anthems, prayers, school songs, or football chants; the reverence for such symbols as flags or queens or presidents; the sharing of a common literature or even a taste for certain television programs—all these things lay down patterns in consciousness that bring us more deeply into correlation with others in our group or nation. Similar, but more private, rituals exist within couples and families. To the extent that we participate in more or fewer of them, we feel more or less at home on the social level of existence, more or less alienated, more or less empty.

The same principle applies to the effect of commitment on our relationship to Nature or spiritual values. To the extent that I expose myself to Nature, involve myself with her—dig in the garden, plant a tree or care for a plant, walk in the mountains—I become more one with her, and thus more "natural" in myself. To the extent that I listen creatively to beautiful music or cultivate a love for beautiful things, I both absorb the essence of this beauty (the relationships that it reveals) into myself and at the same time give the value beauty yet another anchorage in this world. All this has very great implications for the education of children, and offers a physical *raison d'être* for some of the educational principles outlined by Plato in his *Republic.*[18]

Because the basis of a personal commitment is a sense that the other is somehow a part of me, commitment to strangers is more difficult, but not impossible. We do, after all, share the basic nature of our consciousness, a phylogenetic history, and a planetary fate with all other human beings, as well as a weak, underlying quantum correlation. But such commitments require more work and at least a basic introduction to the stranger with whom they are expressed.

It makes little sense to speak of committing myself personally to someone in another country of whose existence I am entirely unaware, but I can certainly feel a commitment to the starving inhabitants of the Sudanese desert or the flood victims of Bangladesh after seeing the harrowing images of their suffering on my television screen.

Such commitments, however, are more transpersonal than interpersonal. They have more to do with spiritual values like love, truth, and beauty—in this case a general love of humanity and a concern for its suffering—than with my immediate, personal relationships to other people. The transforming power of such commitments is not that the distant stranger's existence itself is interwoven with mine, but that my concern for his suffering renews and reinforces my anchorage in transpersonal values. These, in turn, work towards the integration of my individual self with the wider world. By cherishing them, I bring myself into relationship with people and things beyond the reach of more intimate, private, or familiar relationships.

A psychology of the person based on the quantum nature of the self stresses all these relationships and grounds the individual, by virtue of his very nature, in the world beyond the self. For such an individual, for whom commitment to others, to Nature, and to spiritual values is of the essence of his existence, there can be no basis for the narcissistic ills of loneliness, emptiness, alienation, or self-involvement. In the words of the poet John Donne, "No man is an island, entire of itself; every man is a piece of the continent, a part of the main."[19]

To be involved with myself is by the very nature of that self to be involved with others. To be a self in the first place is to be a being within whom all of reality finds expression. As Arthur Miller said about the art forms of Ibsen, Chekhov, and the Greeks:

> What was attractive was that they were forms that allowed, or even demanded, that the individual psychology and society move together in a seamless connection, as it is in life, except we're only half aware of it. The water is in the fish and the fish is in the water. There's no separating the two.[20]

The same applies to the individual psychology and Nature, or to the individual psychology and spiritual values. We cannot separate the meaning of the individual from his involvement with these things.

Equally, a psychology based on the quantum nature of the person carries with it certain basic moral implications, implications that follow *internally* from the very nature of the self—a nature that it shares, at its most basic level, with the whole of reality—and that lay the foundation for a new kind of "natural law" morality.* In this, too, quantum psychology differs radically from the Freudian and existentialist models.

For Sartre, who saw morality as something imposed upon us by a transcendent God, the death of that God meant the noncompulsoriness of His morality. "Indeed, everything is permissible if God does not exist. . . ."[21] I am to be the maker of my own values, the guardian of my own conscience.

---

*The full justification for this statement will become clearer as the remaining chapters unfold.

For Freud, morality is imposed upon us by the cultural superego and its impossible demands are one large source of neurosis:

> It assumes that a man's ego is psychologically capable of anything that is required of it, that his ego has unlimited mastery over his id. This is a mistake. . . . If more is demanded of a man, a revolt will be produced in him or a neurosis, or he will be made unhappy. The commandment "love thy neighbour as thyself" is impossible to fulfil; such an enormous inflation of love can only lower its value.[22]

In consequence Freud and all his followers counseled moral neutrality as a crucial technique in the treatment of patients. Psychoanalysis and psychotherapy were to be "value free" so that the patient could explore his feelings without guilt or inhibition. Any hint of reference to morality gave rise to accusations of "moralizing" on the part of the therapist.

Though few analysts or psychotherapists have intended any such thing, this value-free therapeutic technique has spilled over into the popular mind as a more general excuse for the view that almost any form of behavior is acceptable, or at least condonable, so long as it is "honest" or has its roots in the basic drives or history of the psyche. This has helped to underpin a dangerous moral relativism and a cringing timidity in the face of basic right and wrong.

But in a quantum view of the person, it is impossible *not* to love my neighbor as myself, because my neighbor *is* myself, certainly if we have shared any kind of intimacy. My relationship to him is part of my own self-definition, part of this self that I love if I do love myself.

In a quantum psychology, there are no isolated persons. Individuals do exist, do have an identity, a meaning, and a purpose; but, like particles, each of them is a brief manifestation of a particularity. This particularity is in nonlocal correlation with all other particularities and to some extent interwoven with them.

Everything that each of us does affects all the rest of us, directly and physically. I am my brother's keeper because my brother is a part of me, just as my hand is a part of my body.

If I injure my hand, my whole body hurts. If I injure my consciousness—fill it with malicious or selfish or evil thoughts—I injure the whole nonlocally connected "field" of consciousness. Each of us, because of his integral relationship with others, with Nature, and with the world of values, has the capacity to beatify or to taint the waters of eternity. Each of us therefore carries, as a result of his quantum nature, an awesome moral responsibility. I am responsible for the world because, in the words of the late Krishnamurti, "I am the world." Or, as Jung expressed it:

> If things go wrong in the world, this is because something is wrong with the individual, because something is wrong with me. Therefore, if I am sensible, I shall put myself right first.[23]

Such responsibility alone gives meaning and value to our existence, but to what extent can any one of us fulfill it? If a psychology of commitment and responsibility is to have any value in itself, it must raise the question of human freedom, the question of how far any of us are free to commit ourselves as we choose or to bear the responsibility that is ours by nature. Thus a quantum psychology must embrace some resolution of the problem of free will and lay a foundation for the reality and efficacy of choice.

CHAPTER 12

THE FREE SELF: QUANTUM RESPONSIBILITY

How can life respect determinism on the *without* and yet act in freedom on the *within*? Perhaps we shall understand that better some day.

—PIERRE TEILHARD DE CHARDIN
The Phenomenon of Man

In the British newspapers recently, there was a tremendous outcry about the very lenient suspended sentence given a married man who had raped his eight-year-old stepdaughter while his wife was in the last months of her pregnancy. According to the judge who allowed the man to walk free, his behavior was understandable given the frustration he must have felt from his wife's being temporarily uninterested in normal sexual relations. He was held not to be responsible for his actions.

The cries of shock that followed the judgment demonstrated that public sentiment was clearly at odds with the decision. Most people, or at least the majority of those who spoke out, felt the man should have been able to control his misdirected passion, and that, in consequence, he should bear full responsibility for behavior that was not only illegal, but also morally repugnant. In the end, the court of appeals came to agree, and the man was sent to prison.

The case aroused so much public disquiet because the issues at stake went far beyond the culpability of one loathsome English stepfather. It touched a raw nerve that most of us share concerning the extent of our freedom to act or not act as we choose, and the degree of responsibility we should bear in consequence. Such questions, though they go to the heart of what we are about as human beings, have remained at the edge, or just beyond the grasp, of our best-reasoned arguments.

We certainly *experience* ourselves and others as free, and we order both our feelings and our affairs accordingly. The whole notion of voluntary behavior, or indeed of volition, or the will itself, follows from this, as of course does the concept of spontaneity and a whole panoply of feelings embracing things like awe and wonder, pride and shame.

Whether we think of trivial examples like being free to raise an arm or get up from a chair whenever we like, or more significant decisions like choosing whom to marry or what career to follow, deciding whether to give up smoking, or spend more time with the children, or remain faithful to a commitment, in each of these cases we have a sense that what we do is up to us. In each case we feel that we have chosen or decided freely, and that we must accept responsibility for those choices and decisions. Praise and blame are meted out accordingly.

But such experiences of freedom are, and often have been, at loggerheads with any arguments we might use to defend or justify them. As is true with so much of our experience, it is difficult to argue *rationally* what we know *intuitively.* But in the case of freedom and its attendant responsibility, some kind of rational grounding is terribly important for ordering our social relationships, at the very least.

In any modern society, social relationships are framed in bodies of law. These, in turn, rest upon our best-reasoned arguments about what it is right and wrong to do and to what extent each of us is capable of doing or not doing it. If we cannot *argue* that we are free and responsible beings, capable of deciding between right and wrong and of acting on that decision, we leave ourselves open to the kind of attitude expressed by the rape case judge, or by much of modern sociology and psychiatry in general—attitudes strongly influenced by our modern psychology of the person.

All discussions of human free will—that is, our *internal* freedom, our freedom to have thoughts and make choices, as opposed to any freedom or constraint that might result from outside conditions like political regimes, parental rules, or simple physical ability or lack of it—have been couched in terms of human nature or in terms of humankind's place in the universe. The efficacy or inefficacy of our willing, indeed the question of whether we really have any will at all, is thought to follow from what we are as human beings or from how much power we have over our own actions. Very often in the past, and certainly today, such discussions have lent support to some form of determinism, to the view that our behavior is somehow fixed beyond our control and that freedom of the will is an illusion, an impossibility.

For the ancient Greeks, this determinism was expressed as fate. Helpless before the violent upheavals of Nature and ignorant of their causes, they saw themselves as playthings of often capricious gods, their human actions preordained by forces and plots beyond their ken or control. "What cruel fate has brought me to this bloody deed?" is a cry echoed throughout Greek tragedy, and the concept of tragedy itself rests on the view that certain outcomes are inevitable, whatever we do. Given the hero's character and the situation, they could not be otherwise. As Aristotle says, tragedy arouses feelings of pity and terror,[1] but not of blame.

Equally, in the Christian tradition, many influential thinkers, particularly of the Protestant persuasion, have been convinced that our decisions and actions cannot be otherwise than they are. They cannot in any meaningful sense be free, because there is a divine power or even a divine predestination at work in everything that happens in this world. Owing to God's goodness, His omniscience, or His omnipotence, it follows that everything that *does* happen *had* to happen. As Martin Luther expressed it in the sixteenth century:

> It is, then, fundamentally necessary and wholesome for Christians to know that God . . . foresees, proposes, and does all things according to His own immutable, eternal and infallible will. This bombshell knocks "free will" flat and utterly shatters it.[2]

Other Christian writers allowed a *limited* form of free will, to the extent of saying we are free to seek God's grace or help in ensuring that our choices might be those which are according to His will rather than the Devil's, but without such grace we are wholly lost. As the opening prayer of the Anglican service reads, "God help us, as we are not free to help ourselves."

Most of us today, if we are honest, might admit to vestiges of both the Greek and Christian notions of determinism at the edges of our thinking. We still use phrases like "It was fated to happen," "God must have planned it this way," or "I'll do it, God willing," but these notions of fate or divine predestination have little rational hold over the modern mind.

Today we are enthralled by science and by what it can tell us about the causes of things, including our own behavior. If we doubt our capacity for freedom and responsibility, it is because our science has given us grounds for such doubt. If we are to transcend this doubt, the basis for that, too, will most likely arise from within science.

In fact, modern science has undermined our sense of freedom and responsibility on two fronts, both through its picture of our place in the universe and through the model it has given us for understanding our human nature.

In Newton's dead and silent universe, we conscious human beings have neither any part to play nor any power to still the march of blind and immutable forces. As Bertrand Russell expressed it in a later portion of his deeply pessimistic account of the upwards struggle of the human spirit, "Blind to good and evil, reckless of destruction, omnipotent matter rolls on its relentless way."[3]

For Russell himself, such human impotence gave rise to a defiant faith, but in great numbers of people it leads to a loss of will (depression and despair) or to a ruthless expediency. What does it matter what I do, what choices I make, if in the end it all comes down to dust? My freedom, if I have any, is robbed of all meaning.

This loss is exacerbated by the technological side effects of our science. So many of us depend upon immense and impersonal services for the necessities of life, and we live or work in vast cities or conglomerates where individual choice and action seem to make little if any difference to what goes on around us. This is a recurrent theme of modern literature.

The sense of unfreedom that follows from the impersonal and determinist nature of the classical physics that underpins our science is reflected in the historical determinism of Marx and his followers. Theirs is a determinism imposed from outside by forces beyond our control. But scientific determinism has also got *inside* our modern psychology of the person, denying to us not only the efficacy but also the reality of choice.

In writing his scientific psychology, Freud set out to discover in the human psyche laws and forces that would mirror those in the physics and chemistry of his day. According to Charles Rycroft:

> If, [Freud] argued, all mental activity is the result of unconscious mental forces which are instinctual, biological and physical in origin, then human psychology could be formulated in terms of the interacting

> forces which were in principle quantifiable, without recourse to any vital mental integrating agency, and psychology would become a natural science like physics.[4]

The human psyche, in short, is by its very nature, according to Freud, the bondslave of unconscious forces outside its ken and beyond its control. As one of Freud's followers has commented, this model, if taken as literally as Freud intended, would imply "that all conscious decisions are strictly determined by unconscious forces. . . . that all deciding is an illusion and that consciousness has no function."[5]

Like so much of Freud's conceptual framework, his psychic determinism was not fully realized in his actual practice, and was softened considerably by many of his followers. It contributed, however, to a deterministic bias in much of psychiatry and psychotherapy and left a formative mark on both the academic and the popular minds, raising grave doubts about the human capacity for free and responsible choice.

"Philosophers," says the *Encyclopedia of Philosophy,* "almost entirely agree that if a man's behavior is the effect of a neurosis or inner compulsion over which he has no control and of which he has no knowledge, then in a significant sense he is not morally responsible, and in any case he certainly is not free."[6]

This basic notion that our freedom is limited by the determining power of unconscious instinctual forces soon became the model for more general doubt about autonomy and responsibility.

Our present psychology of the person, as understood by ordinary people as well as by academics, lawyers, and judges, is a curious hodgepodge of deterministic notions either taken directly from science itself or half-digested from a blend of the uses to which Freud and Marx tried to put science. Their original notions that our freedom is hostage to instinct or history have been expanded by sociologists, psychologists, and pundits of all sorts to include claims that our behavior is determined by our environment, by how much money we do or don't have, by our associations, by the media, or even by government policy.

The hooligans in Bernstein's *West Side Story* taunt Officer Krupke with the claim, "We ain't bad, we're deprived. We're sociologically deprived." In Mrs. Thatcher's Britain, everyone is said to be selfish and greedy because she upholds the values of free enterprise and competition. The upshot of all such claims is a diminished expectation of responsible behavior on the part of free individuals. This has undoubted political implications, and has already worked its way into our legal system.

The American lawyer Clarence Darrow was the first to make great use of the notion that criminals can't help being what they are or doing what they do. His brilliant defense orations seldom bothered with any pretence of innocence on the part of his clients, resting instead on claims of their helplessness in the face of forces beyond their control. Many lawyers since have relied on such a defense. The decision of the rape case judge cited at the beginning of this chapter is just one recent example of their success.

In the past few decades, the existing strands of scientific determinism have been reinforced by the extent to which computer technology and the computer model of the brain have excited the imagination. Computers don't make free decisions that carry responsibility, they follow programs. As the computer salesman reminded me when I complained that my new word processor had erased a whole day of my work, "It makes no sense to say that the computer was responsible. You did it. You made some mistake."

But the existence of this "I" who made the mistake exposes the whole deep error of trying to apply the principles of classical scientific determinism to the behavior of human beings. "I" am an active agent, and there are no active agents in classical physics. There are only laws.

If I try to define my "I-ness," my agency, in classical terms, I get into the reductionist trap that I discussed earlier. "I" inevitably fragment into a tangled mess of individual neurons, no one of which can be said to bear responsibility for any actions I might take. There is no place for the buck to stop.

It is only with a quantum model of the person, where I-ness arises out of a coherent, unifying quantum state in the brain, that there can be any one, central "me" who makes or avoids mistakes. This is because the Bose-Einstein condensate that is the physical basis of consciousness generates an electric field that extends over a large region, and any patterns (thoughts, impulses) in the condensate would have a correlated action on many of the brain's neurons, influencing simultaneously their firing potentials and causing them to act as one.

But even given a quantum model of agency, without a similarly nonclassical model of choice, of how the self can exercise its agency, any mistakes this "I" has made must be without blame, without responsibility. In classical terms, there is no scope for the exercise of freedom nor for the consequent bearing of responsibility, no matter what the nature of the self.

In classical physics it is difficult even to define what we mean by *free.* There are models for apparent randomness—the weather, the behavior of a cork bobbing about in a turbulent stream, any of the many examples of unpredictability detailed in James Gleick's *Chaos*[7]—but these are instances where the complexity of a chain of causation is so enormous that we simply cannot fathom it. The causation itself is still there. Therefore, these are not examples of freedom.

In quantum terms, however, it is impossible to define our human being *without* confronting the meaning of freedom. Consciousness, by its very nature as a quantum system, is a thread of freedom running through our lives at every moment.

The physical basis for freedom in any quantum system is quantum indeterminacy, the fact that quantum wave functions can't be "pinned down"—like Schrödinger's cat, who is neither alive nor dead because, as a quantum cat, he is simultaneously both alive *and* dead. That is, he is the joint possibility of becoming either, and *which* possibility gets realized is wholly undetermined. There is no tidy classical law to tell me that looking at him one way will kill him while another kind of look will save him. Such outcomes are strictly a matter of probabilities.

Several people have had the thought that this quantum indeterminacy may bear on the question of human free will,[8] though without a solid quantum model of consciousness itself on which to base it, this insight has remained undeveloped. I think we can hope to get further here.

At the edges of our thought processes, we are all quantum cats, indeterminate quantum wave functions (patterns on the brain's Bose-Einstein condensate) carrying varying and multiple degrees of reality and unreality.

If we reflect gently on the contents of our conscious minds at any moment, we are aware of a dim array of multiple thoughts, of "possible thoughts." These borderline areas of consciousness, the "twilight of the mind" spoken of by some poets, are most accessible just before falling asleep, in states of deep meditation or under the influence of certain drugs, but they are always there, on the edges of any act of concentration. Their reality is fuzzy and their future indeterminate, awaiting some act of realization. Without them there would be no basis for the multiplicity of poetic meanings that distinguish poetry from prose, nor food for fantasy and the imagination.

Freud spoke of these intertwined and multiple images at the edges of consciousness as the "primary process" of mental functioning, our "magical thinking," which serves the purpose of eliminating the tension caused by the conflicting demands of instinct through imaginary wish fulfillment. But he saw the primary process as issuing from a primitive, prelogical stage of our mental development, as something that hinders our adaptation to reality and that must, therefore, be repressed or outgrown. In quantum terms, however, this fuzzy, indeterminate margin of thought is the necessary precondition of *all* thought, reflecting the quantum origin of our thinking. It is the physical basis of our creativity and of our freedom.

Each act of concentration is an act of thought realization. Each of us has had the experience that the process of concentration collapses the wave function of a superimposed array of possible thoughts; though few of us might have expressed it this way before being introduced to a quantum vocabulary. By focusing on any one thought, we make that one into a classical reality while the others disappear like so many shadows in the night.

Thus each act of concentration expresses a mini form of choice, a mini form of freedom. Nothing determines upon which one of an array of "possible thoughts" I will focus, because the "I" that focuses is itself an indeterminate quantum wave function, but through the act of focusing a choice is made. Through the act of observing Schrödinger's cat, I kill or save him; through the act of observing my own consciousness, I actualize or lose some of my possible thoughts.

A trivial example may help to make the indeterminate nature of choice more concrete. If I feel tense after sitting here at my desk for several hours writing this book, I may find myself gazing off into space, my head filled with images of smashing my computer, going for a walk, jumping up and down, riding my bicycle across the meadow, or continuing to sit here until I turn to stone.

Very briefly, I see all these images at once, *live* all of them at once, like the quantum hussy who made a home with each of her lovers simultaneously. But my physical discomfort will spur me to concentrate, and when I do so, I will—by that very act of concentration—

choose one possible source of relief for my tension and act on that choice. A choice *is,* in these terms, nothing but an act of concentration that collapses the wave function of possible thought.

But no one can say that *that* specific choice was determined by my discomfort. Any one of my choices would have relieved it. The discomfort only necessitated *some* choice. The choice itself was free.

This self-reflective capacity of thought to observe itself and thus, through concentration, to collapse its own wave function rests on the physics of at least some Bose-Einstein condensates (including those that are the physical basis of our consciousness), on the different physical properties displayed by such quantum systems when they are in a low-energy state or a high-energy state.

In a low-energy state, Bose-Einstein condensates display the familiar quantum superposition effects of multiple possibilities, experienced by us as the blurry images of our dream life, the Gothic twilight of the imagination. In a high-energy state, these condensates behave almost classically, losing their quantum superposition effects.

This switchover mechanism, taking the condensate from low-energy quantum properties to high-energy classical properties, was first illustrated in Josephson Junctions, the superconducting loops that won British physicist Brian Josephson his Nobel Prize in 1973.[9]

Because they can bring together in one macroscopic (large-scale) physical unit the properties of both quantum and classical systems, Josephson Junctions have raised some hope for the invention of quantum computers,[10] computers that could combine the advantages of quantum superposition effects (chiefly, the freedom to choose simultaneously from a range of possibilities) with classical computer logic. But superconducting Bose-Einstein condensates work only at extremely low temperatures, so the technology of employing Josephson's discovery in any actual computers, classical or quantum, has so far proven too expensive to be feasible.

In the human brain however, where Fröhlich-style Bose-Einstein condensates function at body temperature, there is no such problem. The brain is, therefore, a successful example of a "quantum computer" employing the basic physics of Josephson Junctions.* Saying this, how-

*An aside to those who want to know more about the physics of this: In a standard superconducting ring, the circulating electric current has a definite value. But in a Josephson Junction, which is a ring containing a weak link, this link weakens the energy value of the condensate and

ever, has vastly different implications for understanding ourselves as humans than for any comparison between ourselves and classical computers, for reasons that I hope are clear in everything being said about the nature of ourselves as quantum systems.

In our conscious system, the act of concentration is the process by which energy is pumped into the brain. We are all aware that when our energy reserves are low, we find concentration difficult. But when we do have the energy for concentration, channeling this energy into the brain has the effect of switching the brain's Bose-Einstein condensate from a low-energy quantum state to a high-energy near-classical state, and thus of switching our thought processes from the blurry images of possible thought to the more structured, classical detail of concentrated thought.

On a quantum view of consciousness, then, we have both a basic definition of choice and a basic understanding of the physics that makes choice possible. Any choice, itself, is simply the collapse of the quantum wave function of possible thought into one definite thought, and the physics by which this happens is the switchover of the brain's Bose-Einstein condensate from a many-possibilitied quantum state to a more definite near-classical state. All such choices are necessarily free because of the brain's essential quantum indeterminacy—an indeterminacy that exists both in its quantum system and in the firing responses of individual neurons to stimulation.[11]

But this bare-bones model of quantum choice still leaves unanswered all the most interesting questions. How and why, for instance, do I actually make the choices that I do, and if I am free to make *any* choice, why do I so often make what are clearly bad choices, bad for myself or for others? To what extent can I control these apparently indeterminate quantum choices—control my own freedom, in other words—and thus to what extent does my freedom make me responsible for my choices?

The commonsense answer to many of these questions, if I believe

---

allows quantum superposition values to occur. Something similar happens in a Fröhlich system. One definite state of consciousness, I am suggesting, corresponds to a complicated circulatory motion, related to phase differences among the oscillating dipoles, around vortices in the condensate. This circulation pattern has a definite value only when there is a high-energy input (during concentration), and becomes a quantum superposition in a low-energy state. (See I. N. Marshall, "Excitations of a Bose-Einstein Condensate," forthcoming.)

in freedom at all, if I am not a determinist, is that I am a rational creature, that I have the capacity for the logical analysis of situations and for reflection on the likely outcome of any choices that I make. Both my freedom and my responsibility are thought to follow from such capacities—that is why we so often deny that animals have free will or that children should be held responsible for their actions.

But this notion that choice and reason are necessarily linked in the decisions of a free person is somewhat overstated and blinds us to the true nature of choice and freedom in the quantum self. Quantum freedom is a far more terrible thing than our faith in the power of reason would have us believe.

Take, for example, making the choice to give up smoking. All my capacities for reason tell me that smoking is bad for me and most likely for all those around me. If I listen to reason, I will undoubtedly decide that I *should* give up smoking. I may even convince myself that I am acting on this decision by promising to give it up "tomorrow," or by subscribing to some gimmick like hypnosis or acupuncture. But the effects of the gimmicks are short-lived and tomorrow is a long time in coming. I continue to act against reason, to *choose* against reason every time I light another cigarette.

But one day I *do* give it up. One morning, when I have no reason to suspect such a thing would happen, I reach for my packet of cigarettes only to throw it down again. I have chosen to quit. I have actually made my choice and acted upon it. But why?

In quantum terms, that "why?" has no definite answer. All definite answers—all logic and reason—are classical structures. They arise just at the point when the wave function of thought collapses, that is, *after* the moment of choice. Our logic does not make our choices—that is a determinist way of thinking.

Rather, our choices, our free and undetermined choices, which are associated with a similarly superposed set of reasons linked to those choices, give rise to our logic. In making a choice, we also make a reason for that choice, a reason that our logic then uses to explain the choice. But some *other* choice would have been associated with some *other* reason, which would have served equally well logic's thirst for explanation.

I will say that I gave up smoking *because* I knew it was bad for my health. But equally, had I failed to give it up, I would have said that this was *because* I was too weak-willed, or *because* I needed it to relieve my tension. These 'becauses" that I use to explain my choice tell something about me as a person, but they don't determine the choice itself.

Some psychoanalysts and psychotherapists believe that the true value of their work arises not from any dubious ability to assign causes to their patients' behavior, as Freud argued, but rather to discover the meaning of the behavior—to discover what the making of certain choices tells us about ourselves and what we value.[12] The choice to give up smoking tells me that I value my health and longevity, and beyond that perhaps that I am the sort of person who can resist immediate temptation for the sake of distant gain; a choice not to give up smoking might have indicated that I valued immediate though temporary pleasures above long-term benefits.

But whatever the meaning of my choice and what it tells about me, the choice itself preceded all "becauses." It was made in a terrible moment of freedom, in what Kierkegaard would call a "leap of faith."

Nonetheless it was I who made the choice, I who, in some strange quantum dialogue between the indeterminate quantum wave function that is myself and the indeterminate wave function of my possible choices, did decide to quit. And that choice was the responsibility of no one and nothing but myself.

That is the dreadful burden of freedom, that it makes us responsible for choices over which we have no full, conscious control. It places us in the line of fire, in the midst of a situation that is all fuzzy and indeterminate at the edges, and then tells us that, in "fear and trembling" as Kierkegaard would say, we must stand up and be counted.

But, we want to cry out, can life and freedom really be that terrible, that fraught with awesome choice for which we must bear responsibility, and yet which arises from within a realm of the self that apparently answers to no one? Is there nothing I can do to control my freedom, to rein it in just a bit?

For Sartre, whose radical existentialist freedom was founded on a denial of both human nature and essence as well as the efficacy of any outside determining forces like rules and values, the answer was a firm no. "I *am* my freedom," cries Orestes in *The Flies:*

> Suddenly, out of the blue, freedom crashed down on me and swept me off my feet. . . . I was like a man who's lost his shadow. And there was nothing left in heaven, no right or wrong, nor anyone to give me

> orders. . . . Foreign to myself—I know it. Outside nature, against nature, without excuse, beyond remedy, except what remedy I find within myself . . . I am alone, alone. Alone until I die.[13]

Sartre himself is forced to conclude that this position means "human life begins on the far side of despair,"[14] and leaves unanswered the question of what grounds there might possibly be for finding any remedy within the necessarily empty existentialist self. Placing any value at all on such radical freedom is just another expression of the narcissistic wish that the self be grounded entirely on itself, with the necessarily consequent alienation from both self and others. As the Canadian philosopher Charles Taylor sees:

> The subject of radical choice is another avatar of the recurrent figure which our civilization aspires to realize, the disembodied ego, the subject who can objectify all being, including his own, and choose in radical freedom. But this promised total self-possession would in fact be the most total self-loss.[15]

The narcissist's dream is its own nightmare.

In bold contrast the quantum self is neither empty nor alone, nor is it radically free in Sartre's sense of the word. Neither, for that matter, are quantum processes. The collapse of a quantum wave function is not random, not wholly without a "sense of direction"—not, to use Sartre's vocabulary, wholly contingent. Any collapse is a matter of probability, and some outcomes of a collapse are more probable than others. For human quantum systems such as ourselves, the extent to which we can weigh those probabilities is the extent to which we can exercise some control over our freedom.

In quantum processes, a probability that something will happen is associated with the amount of energy required to make it happen. If an electron can move to one energy shell within the atom with very little expenditure of energy, and to another at very great expenditure of energy, the probability is very high that it will make the low-energy transition. It is free to make any transition, nothing is determined, but it is most likely to take the easy option. So it is with us, though because we are very much more complex characters than electrons, the factors that influence the energy requirements of our various choices are also more complex.

As a quantum person I have both a nature and an essence. I have a body, genetic dispositions, experiences and reflections on those experiences, a character, and I am defined largely in terms of the relationships I establish with others. All these qualities have an impact on my quantum memory, on that indeterminate meeting point between the self that I am and the self that I am becoming—on that point where my choices are made. And the nature of that impact is that it influences the probabilities of my choices. The whole history and makeup of my being increase the probability that I will make some choices, and decrease the probability that I will make others.

As quantum selves, we make ourselves as we go along, weave the fabric of our being through our ongoing dialogue with our own pasts, with our experience, with the environment, and with others. An important part of that dialogue is the reasons that we assign to the various choices that we might make and how those reasons fit into the whole context of our lives and what we value. Thus, while reasons themselves don't *determine* the choices we make, they do play a crucial role in making some choices more likely than others. The particular reasons that we link to an array of possible choices influence the probability of making any particular choice.

The reason linked to the possible choice of giving up smoking is that it will prolong my life; the reason linked to not giving up is that I enjoy it. But given the association of these reasons with those choices, it is more *probable* that I will choose to quit. The association between reason and choice makes the right choices easier, less energy demanding; it tips the balance, but it does not guarantee the desired outcome.

Through the whole process of our living and our thinking and our relating, we are reinforcing or shifting the probabilities that our choices will have a particular outcome. We are loading the quantum dice and channeling the direction of our freedom. Each choice that I do make has an influence on the next choice that I will make, because it increases or decreases the probability of that choice. None of my choices, however small, is without some significance for the rest of my life.

Obviously such channeling has greater efficacy as an individual's character and reasoning capacities mature, and we are right not to hold children or the mentally defective as responsible for their actions as we do normal adults. Their freedom is as real as that of the mature adult, but its outcome is likely to be far more random, or far more weighted by genetic disposition or chemical imbalances in the brain.

The effect of our life-styles and all our past choices on weighting the probabilities of our future choices also lends a certain *limited* truth to sociological or psychological claims that our backgrounds, our surroundings, or our associations influence our choices. But that is very different from saying that they *determine* them. We are always free to choose against the weight of probability, to make more energy-demanding choices, and this freedom makes us responsible.

The stories of people who have risen above their backgrounds or circumstances to do surprising or great things inspire us just because they remind us that we, too, *could* act against the odds, that the responsibility for this lies with no one but ourselves. And, as we see so often, that realization often changes the odds. That is why the examples of local heroes often transform the lives of many others in a disadvantaged or downtrodden community. Their having made the hard choice makes that same choice easier for others.

The physics of this lies in the quantum interconnectedness of our consciousness and reflects my earlier statement that everything that each of us does affects all the rest of us, directly and physically. If one of us blazes a trail, it is more likely that others will take the same path.

On the whole, the quantum nature of our consciousness makes it tempting for us to make choices that require the least expenditure of energy, the least amount of concentration. And that is why we are by nature creatures of habit or imitation.

Habit is a kind of free ride; it requires very little mental work. Once I have done something one way, made some particular sort of choice, it is very much easier to do the same thing again, and therefore the probability is greater that I will do so. That is why I should use all my best reasoning faculties to assess the value of the habits I am adopting or the qualities of those whom I am imitating. The original choice leading to a habit may cost me little, but if I later wish to break the habit, the task may be of Herculean proportions.

In one sense, any habit is an escape clause for the lazy and the timorous. It saves us energy and at the same time it relieves us of the burden of our freedom. Once something has become a habit, the probability that we will repeat it is so great that there is almost no element of choice left in the situation. Thus when I act out of habit,

I do not act out of my freedom, nor do I exercise my creativity. Being a low-energy activity, habit pumps very little energy into my brain. It collapses few wave functions, as it were. That is why it is so uncreative and why creatures of habit experience so little psychic growth.

But perhaps the habitual is necessary in a great many areas of our living. Perhaps we simply haven't sufficient quantities of physical energy to live at the edges of our freedom with every decision and action that we take, and perhaps that is why the very quantum nature of our consciousness tempts us towards the habitual. The cultivation of habit may free us for more creative living where it really counts.

The same applies to letting our actions follow from conformity with accepted codes of behavior or from adherence to strictly defined codes of duty. The initial choice to follow these codes requires some concentration, though not much if we are already partially defined in terms of the social mores and relationships that support them. But once that choice is made, we can continue to live in a way that so loads the probabilities against any contrary form of behavior that it would require the qualities of a hero to muster the energy to do so.

None of us can be heroes at every moment of our lives, and so long as the accepted mores or the codes of duty to which we subscribe are basically decent, the necessity for individual heroics can be avoided without damage to oneself or others. If our adherence to accepted mores follows from a commitment (which is a decision renewed, with energy, time and again), rather than from mere habit, then conformity itself may be a creative way of living. It helps to sustain a culture and a life-style.

But because of our essential freedom, and because of the responsibility imposed upon us by that freedom, any one of us may at any time be called upon to be a hero. When our habits can be seen to be bad for us or for others, or when our loyalty to duty involves us in behavior that we know is morally wrong, we are compelled to become heroes, compelled to make the effort to act against the weight of probability.

We have a moral imperative to use our freedom, to live at the fearsome edge of our consciousness when called upon to do so, because it is our very nature as conscious beings to be free, and in quantum terms the natural and the moral go hand in hand. Why this is so will become more clear when I have discussed ourselves in relation to the material world and the nature of the quantum vacuum itself.

The exercise of this freedom lies at the heart of what we are about as quantum individuals, and we rightly blame those who always shirk it in the name of duty or habit or social conditioning. We blame them for not making an effort (spending some energy), for habitually shirking the burden and the responsibility that is ours by nature, and thereby opting out of the creativity to which our freedom gives rise. That creativity, which I shall discuss next, is the key to understanding why we conscious human beings are in the universe at all.

CHAPTER 13

THE CREATIVE SELF: OURSELVES AS COAUTHORS OF THE WORLD

> We are the bees of the invisible. We madly gather the honey of the visible to store it in the great golden hive of the invisible.
>
> —Rainer Maria Rilke
> Letter to Hulewicz

The essential creativity of human beings runs as a theme throughout our history and culture. We see ourselves as "man the maker," and in modern scientific terms date the origin of our species back to the day that the first man made the first tool. We feel that our creativity somehow separates us from the beasts and defines our humanity.

In religious terms, our creativity has sometimes been seen as the *reason* for our humanity, as the *raison d'être* of human existence. This theme appears, for instance, in the Jewish mystical tradition, which argues that God made man because He needed a partner in creation,[1] and in the philosophy of Henri Bergson, who believed that the whole purpose of the evolutionary process was God's "undertaking to create creators."[2] It is also, of course, an idea running through so much of Rilke's poetry.

"Earth! Invisible!" he writes in *The Duino Elegies.* "What is your urgent command, if not transformation?"[3] And it is to us that this command is given, to us "bees of the invisible" who through our living both realize and transform the earth's silent potential. Something about our human nature is such that creativity lies at the heart of what we are about.

We certainly feel this about ourselves in small ways as we go about our daily lives, and if we reflect on our behavior, we can often see that a "creative urge" motivates much of it. Simple things like children's first paintings or first attempts at piling one wooden brick on top of another; the later wish to build models, carve soap, make clay pots and baskets; the adult penchant for do-it-yourself; and the wish to decorate both ourselves and our homes are all basic expressions of the same drive that motivates others to write poems or symphonies or to articulate new religious views.

More basically still, we recognize that there is something creative in meeting any new challenge, making any new relationship, charting any new path. These activities, like their more artistic counterparts, stimulate us and cause us to grow—they create something *within* us. When there is no obvious outlet for the drive that motivates them, we feel bored or stale, or even that our humanity itself has been diminished—hence references to the "dehumanizing effects" of routine factory work or bureaucratic systems that leave no room for initiative. To relieve such boredom, we go out of our way to invent challenges, like sports and games, daredevil stunts, or even taunting the police and petty crime. All are expressions, if sometimes distorted expressions, of a deep *need* to be creative.

Yet this same creativity, which seems to define so much of what we are, remains itself deeply mysterious. In any terms we are accustomed to using, it is difficult to say exactly what creativity *is*, to say what is happening when a child makes a clay pot or a person freely responds to some challenge. We feel intuitively that each of these is different from a machine turning out dozens of identical dishes or a computer following programmed options. We feel that neither machines nor computers are creative—but why? What is it about the handmade pot or the human response that distinguishes both from their mechanistic counterparts? Ironically, the beginnings of an answer may lie in the extent to which human beings are not unique.

Important developments in the science of the past few decades have shown that at least some of the creativity we associate with ourselves extends, at a very elementary level, to all life. Something about the structure of living systems themselves—from the humblest yeast cell to a complex human being—is such that their very existence creates a special kind of order. This order is what mediates between the dull monotony of things just "lying around" or carrying on in some routine or determined way and, at the opposite extreme, those things affected by the disturbing turbulence of chaos.

This living order somehow manages to get round the Second Law of Thermodynamics, which claims that everything in the universe is running down, or falling into disorder (the law of entropy). Its discovery lies at the heart of Ilya Prigogine's Nobel Prize–winning work on dissipative, or open, systems[4]—of which living systems are one important kind.* It is also linked to Herbert Fröhlich's discovery that there is quantum coherence (quantum order) in living cells. A Fröhlich system *is* a living quantum system.

The kind of order created by living systems is not the order of tidying-up operations. Weary housewives are right to complain that clearing up toys and dishes all day exercises little of their creativity. The creativity of living systems—at least that which has its roots in their quantum coherence—arises from their ability to create the kind of order that gives rise to relational wholes, systems that are greater than the sum of their parts, and to do this spontaneously whenever a critical level of complexity is reached. Prigogine calls them self-organizing systems. They are a law unto themselves.

The aliveness of an amoeba is more than the combination of some hydrocarbon atoms mixed up with some salt water; that of a human body is more than heart plus lungs plus kidneys and so on. It is the way these constituents are drawn together in a coherent living system that creates their aliveness, and that coherence simply can't be broken down into so many building blocks. It is a new thing in itself, a new and ordered whole that has grown out of the peculiar relation linking those building blocks and that utterly transforms both their meaning and their physical potential. It is a distinctively quantum phenomenon.

This capacity of living systems spontaneously (freely) to make ordered, relational wholes is, I think, the basis of all creativity,† and to this extent creativity is a capacity we share with every amoeba and earthworm. But by extending these insights to consciousness itself, to the source of our mental, psychological, and spiritual life—which we

---

*Others are convection patterns in air and water, and whirlpools in turbulent streams. The essence of these systems is that they are not isolated—energy or matter flows through them, and as it does so, it gets organized into a particular pattern that is itself both stable and dynamic.

†The physics of this rests on two pillars: the ability of any self-organizing Prigogine system (classical or quantum, living or nonliving) to create a kind of order where there was none before, and the unique ability of quantum systems to overlap and share an identity (thus drawing themselves into new and larger wholes). Without the order, quantum relational holism builds nothing in particular; without the relational holism, self-ordering systems create nothing new. But together they give us the living world. Every living cell is a special, Fröhlich-style, quantum subclass within Prigogine self-organizing dissipative systems.

can do by seeing consciousness as a Fröhlich-style Bose-Einstein condensate in the brain—we begin to see the origins of higher forms of creativity, of those forms that we recognize and appreciate as specifically human. Equally we begin to understand what is happening when a child makes a clay pot or a human being responds to a challenge.

When the child makes his pot, he gives shape and meaning to something that has never existed before. He draws together a hitherto-unexpressed idea in himself and a so-far-unmolded collection of clay molecules and transforms both into a new thing that *is* the relation of his idea to that clay. More than that, it is the relation of himself, of his evolving sense of beauty, to that clay.

The child's creative act has given rise to a new thing (the pot), to a new articulation of his sense of beauty, and to an incarnation of the relationship between the child, his sense of beauty, and the clay pot. Thus in making the pot, the child has also made (some new aspect of) himself and some small portion of his world—his relationship to things. By contrasting the physics of this with the mechanics by which a machine turns out a dozen identical dishes, we can fully appreciate why the one is creative and the other not, and why our creativity lies so close to the meaning of our existence.

To begin with, the origins of the child's clay pot lie in the free dialogue within his brain's quantum system. This dialogue is between a superposition of many possible clay pots, all existing simultaneously as possibilities like so many alive/dead cats, and a superposition of many possible notions of beauty. Neither the pot nor the sense of beauty with which it eventually will be associated has yet been realized. Their wave functions have not collapsed.

This situation already differs from the machine's manufacture of a dozen identical dishes in that each of the machine-made dishes follows after a fixed blueprint. If we look at the blueprint, we know in advance what kind of dish will emerge from the machine and know what notion of "dishiness" has been built into its mechanism. There is nothing left to be resolved. (All the wave functions have collapsed.) By contrast, if we could look at the child's superposition of pots or superposition of his different ideas about beauty without disturbing them, we could see that everything is yet to happen (Figure 13.1).

As the child concentrates on making his pot, he pumps energy into his brain and alters its quantum state. The wave functions of his superposed possible thoughts begin to collapse and both clay pot and sense of beauty emerge. Neither determined the other, and neither was

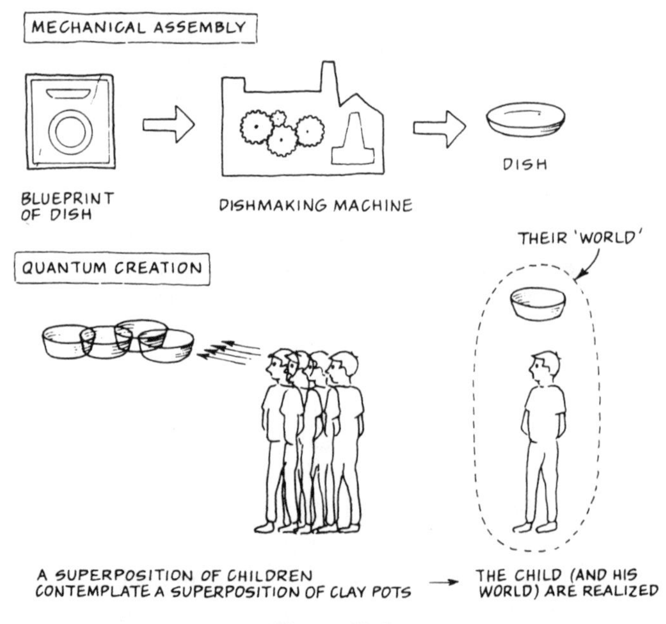

Figure 13.1

determined in itself. There were many *possible* pots the child might have made and many possible ideas of beauty with which these might have been associated. Both the *actual* pot and its incarnated sense of beauty originated in the child's freedom—in the quantum indeterminacy that underlies all his thought processes and decisions.

The whole process by which the child makes his pot is a chain of free decisions. First comes the idea to make a pot in the first place (instead of a clay man or an airplane or whatever), then the idea to make this particular sort of pot, then the decision to press the clay a bit here and to round it off a bit there, and so on.

As these decisions unfold, the child slowly discovers both his pot and that within himself which likes to make such things, but this discovery is a *creative discovery.* It is precisely through this discovery process that the child makes both the pot and himself (his sense of beauty).* His unfolding discovery literally snatches the pot and this aspect of himself from the shadowy realms of possibility and actualizes them. His cre-

*For my original concept of a creative discovery, I am indebted to my old teacher, American philosopher Samuel Todes. His use of the term is similar to that being described here, though I think it gains added meaning in association with quantum theory.

ation has acted as midwife to the birth of a small bit of new reality.

An intuitive understanding of the relationship between a child's creative play and the embodiment of his own potential person lay at the heart of Melanie Klein's use of play interpretation in her child psychoanalysis. She believed that through his play the child both discovered himself and became *more* himself. The same theory underpins the use of art and music therapy for adults—the belief that through weaving a basket, making a painting, or writing a song the patient can bring something of himself into being. This creative self-discovery has its roots in the physics of consciousness, which is utterly different from the physics of a machine.

Both the mutually creative aspect of the child's clay pot and the freedom in which it originated are foreign to the process by which the machine turns out its dishes. Behind every machine there is a creative human being reenacting the drama of the child and his pot, but for the machine itself, the exact style, color, and size of the dish it will turn out are wholly determined by the mechanism of its construction. The machine has no free will.

Random patterns generated by a machine—like the sets of pictures displayed on a slot machine—are not determined. The sense of freedom associated with randomness, however, is wholly different from that associated with purpose or intention, and none of its products would be truly creative.

Equally, the dish-making machine's mechanism is determined by the original blueprint for the dishes that are to be manufactured. Neither the machine nor the design of its dishes is altered as a result of the manufacturing process. They are like two Newtonian billiard balls bouncing off one another—they meet, but each is left unchanged by the encounter.

The essential difference between the child and the machine that makes the one creative and the other not is that the child is in a constant, mutually creative dialogue with his environment and the machine is not. As Wordsworth expressed it long ago, his mind is "creator and receiver both, working but in alliance with the works which it beholds."[5]

In following his natural quantum impulse* to make an ordered

---

*More accurately, his natural Prigogine-style quantum impulse—only complex, self-ordering (Prigogine) quantum systems would display a natural and irreversible direction of evolution. (See Prigogine and Isabelle Stengers, 1984, pp. 297–310.) All Fröhlich systems have this quality.

relational whole from the data of his experience—his quantum impulse to integrate himself—the child *ipso facto* draws together an object (his clay pot) and a world (his relation to the clay pot, its meaning for him and for others) that were never there before. Child, object and world are all co-realized through the free and undetermined collapse of many possible children, objects, and worlds in the child's mind.

All quantum systems (most especially boson systems like ourselves) share this mechanism for creative self-discovery through a dialogue with their environment. At its most elementary level, this dialogue is evident every time a photon travels through one slit or two, or manifests itself as a wave or a particle in response to the presence of a detection screen or a photomultiplier tube.

At the more complex level of living systems (*ordered* quantum systems), there is recent evidence to suggest that biological evolution itself may be, in fact, *responsive* evolution."[6] It may possibly be a quantum dialogue between the creature and its environment, a dialogue that has the capacity to elicit and realize one of many *possible* directions of evolution (mutations) latent in the DNA code. The likelihood of this is strengthened by recent evidence for quantum coherence within DNA itself.[7]

All living systems evolve, and to that extent have a kind of creativity built into their structural development. There is, as Ilya Prigogine says, an arrow of time in living systems pointing towards more and greater complexity: "Time is construction."[8] Or, as German physicist Fritz Popp expresses it, "The coherent state is like a white paper which is always asking us to write upon it."[9]

We have a physical *impulse* to be creative that follows from the physics of living systems. But beyond this structural creativity, a quantum interpretation of consciousness shows us how there can be *behaviorial* creativity—that of the child and his clay pot—which also extends from human beings very far down the evolutionary scale to very simple creatures.

Even earthworms display a primitive tendency to integrate their sensory data and slowly to evolve a life-style, a world. They respond to stimuli in their environment, they learn that certain responses to these stimuli generate pleasure (roughly speaking) and others pain, and they learn to behave accordingly. Some have even been taught not to move when exposed to light, despite an instinct to do so, and others have even learned to run simple mazes.[10]

It may be wrong to say that earthworms are "purposive" in their world construction, or even perhaps to say that they "choose" one form of behavior over another. Purposes and choices are human categories that derive from specifically human (or at least higher-animal) capacities. But it is important to appreciate that elementary behaviorial creativity is available even to very simple living systems as a consequence of their quantum capacity to be in dialogue with and to integrate data from the surrounding world (their very elementary unity of consciousness).

What makes human creativity so much more impressive and awe-inspiring than that of the earthworm is not that it is different in principle, but that it is different in kind, or degree. Our creativity issues from a living system that is infinitely more complex and that has the capacity for rational analysis and self-reflection (self-consciousness). The rational analysis arises from the extraordinarily complex data-processing capacities of the brain's computing system (all those neurons and their connections); the capacity for world integration and self-reflection arises from the impressively large Bose-Einstein condensate of the brain's quantum system and the dynamics of quantum memory that rest upon it.

Without the Bose-Einstein condensate and its capacity to support the unity of consciousness and the construction of relational wholes, we would be nothing but ambulatory computers. Without the computing system to generate excitations (patterns) on that condensate, we would be no more than laser beams.[11] But the two together, the capacity for logical structure and the capacity for integration and creative self-reflection in dialogue with the environment, give us the elaborate creativity that accounts for the human world.

In many simple ways, this creativity happens within each of us in the normal course of leading our daily lives. How it does so, and how the creative living that each of us experiences underpins the larger-scale creativity of our culture, can be illustrated by looking at the physics of how a person responds to a moral challenge and how, through that response, he creates both himself and his moral world.

The very concept of "a moral challenge" (morality) is already an ordered relational whole created in response to our need for an integrated picture of appropriate social behavior. It is an attempt to bring order into the potential chaos that might arise from the very wide range of possible behavior that could result from the actions of complex and essentially free human beings.

In struggling to bring about this order, we give birth to ourselves and to our morality, to a new dimension of consciousness that both expresses and transcends the behaviorial decisions of individual members of any society or group. Each of us helps to write the morality by which we all shall live. This is particularly so at times of moral crisis or moral challenge.

Imagine, for example, that I have become bored with my husband and am tempted to have an affair with another man. This temptation throws me back on my freedom and forces me to choose between my husband and my lover, or at least between fidelity to my husband and an affair with my lover. But the necessity of this choice calls my whole world into question—the whole gestalt of who I am and what I value. It presents me with a significant moral challenge.

Because of the quantum nature of choice—that it is a free choice between a multitude of superposed possibilities that exist simultaneously (crudely, in this case, the possibility of the affair and the possibility of fidelity)—the temptation itself already has an influence on myself and my world. The temptation has evoked the possibility of infidelity, and while that possibility is real, its effects can be felt. In this case, I am likely to be impatient with or unloving towards my husband—in general, halfhearted in my relationship with him. *Half-hearted* is an apt psychological equivalent of quantum superposition—it is neither here nor there.

This is the reality of Saint Paul's warning that the sin is in the thought. The physics of it rests on the physics of virtual transitions—those quantum "trial runs" that I discussed in Chapter 2 and illustrated with the multiple and simultaneous affairs of the quantum hussy. In her case, the outcome of a virtual transition may have been a child by one of her virtual unions. In mine, it is likely to be a row with my husband that will have a lasting effect on our relationship whatever my ultimate, actual choice.

The same applies to filling the mind with any sort of suggestion—whether advertising slogans or ugly images from pornographic videos. Even if we don't act on the temptations they raise, the temptations themselves affect the general health of our individual and shared consciousness. Remember David Bohm's remark that "a great many physical processes are the result of these so-called *virtual* transitions."[12]

In finally deciding between my two choices, I am deciding between two selves that I might become and between the different worlds they might occupy. The choice is free; nothing determines it. Though the character I have built and the life that I have led up till then will weight the probability of my choosing one or the other, I can, and often do, act "out of character."

Equally, any arguments that I may put to myself about the wisdom of one choice rather than the other don't actually determine the choice itself. I don't say to myself that I value my marriage and all the commitments deriving from it and thus choose to remain faithful to my husband, or that I value romance and spontaneity and thus choose the affair with my lover. These are causal explanations, which simply don't accord with my freedom. (It is not my logic which creates my choices, but my choices which create my logic.)

Rather, it is only when I have made my decision that I discover what I value, what really matters to me, and what kind of person that I am. But this is a *creative* discovery—it is precisely through articulating the reasons for my choice that I become the sort of person who would make that choice. As Charles Taylor expresses it in his discussion of the radical reevaluation that accompanies free moral agency:

> Articulations are not simply descriptions. . . . On the contrary, articulations are attempts to formulate what is initially inchoate, or confused, or badly formulated. But this kind of formulation or reformulation does not leave its object unchanged. To give a certain articulation is to shape our sense of what we desire or what we hold important in a certain way.[13]

It is to shape ourselves.

This creation of the self through an articulation of the values that accompanied a given choice is reminiscent of the "backward causation" illustrated in John Archibald Wheeler's delayed-choice experiment (Chapter 3), and probably rests on the same basic physics. In that experiment, a photon has to "choose" between realizing itself as a wave or a particle,* between traveling through one slit of a two-slit apparatus

---

*I very much doubt that photons actually make choices. My anthropomorphic language here is simply intended to draw out the suggestive analogies between their transitions from possibility to actuality and our own.

or through both slits. If it has chosen to be a particle it will travel through one slit; if to be a wave, through both slits. It "articulates" that decision when it strikes either the detecting screen or the photomultiplier tube but, according to Wheeler, it is only when that articulation is made that we can look back into the photon's history and say how many slits it went through. The articulation of that choice created both the photon's character and its history.

Similarly, it is only as I articulate the values that led to my choosing my husband or my lover that I become the person who has those values, the person who has the character to sustain such values and the history that led to them. But in thus creatively discovering myself, I also creatively discover the values that I cherish. I bring these values into the world, or I reincarnate, and thus give new life and meaning to, old values. In doing so, I help to create my world and that of others.

If I choose fidelity (and all its attendant values), my choice acts to weight the probability that others will make the same choice. My self is interwoven and in nonlocal correlation with others in my group or society, and the moral decisions that I make resonate through the world that we share, the world that together we create. If I choose to break my marriage vows, I make it that much more likely that others will do so, that more families will be split, that social instability will increase, and so on.

There is no end to the chain of influence that follows from my decision. I am responsible for the world because I help to make the world. As Jung expressed it when discussing the many crises of modern life and the self-reflection (articulation) with which we should respond:

> In the last analysis, the essential thing is the life of the individual. This alone makes history, here alone do the great transformations just take place, and the whole future, the whole history of the world, ultimately spring as a gigantic summation from these hidden sources in individuals. In our most private and most subjective lives we are not only the passive witnesses of our age, and its sufferers, but also its makers. We make our own epoch.[14]

But these values that we create "in our most private and subjective lives," and through which we create a world, are not themselves subjective. They are not, as Sartre would claim, without foundation beyond the self. In Sartre's version of a self-made morality, it is I, alone and in dread before the awesome fact of my freedom, who create and maintain values in being.

> Nothing can assure me against myself; cut off from the world and my essence by the nothing that I *am,* I have to realise the meaning of the world and of my essence: I decide it, alone, unjustifiable, and without excuse.[15]

But in a quantum world, creation is never *ex nihilo.* The values that I create are not *self-*made (and especially not those made by a self that is nothing). My choices are not made in lonely isolation, nor are the values that emerge from them merely capricious, or relative to my situation. Rather, their creation is evoked by the free dialogue between the self that I am now and my world as it is now—my world of others in relation to whom myself is defined and the world of the human nature that I share with them. As American philosopher Lawrence Cahoone expresses it:

> The subjectivist anti-culture cannot understand that human beings create, think, and become individuated, independent creatures only within and through a context of meaningful relations to other human beings and to non-human beings.[16]

Our relation to the selves and values (worlds) that we create is one of coauthorship. We bring our world and ourselves into being through a shared, creative response to the world and to each other. This introduces a new quantum concept of "shared subjectivity," a subjectivity that is in dialogue with the world and that, through that dialogue, gives rise to objectivity. It is the relationship between the observer and the observed, translated from the physics laboratory into the moral sphere through the quantum nature of our consciousness. It is what Ilya Prigogine calls "a concept of knowledge as both objective and participatory."[17]

Because our human nature is such that we *are* our relationships, and our world is such that we create it together through our common humanity, there is an underlying physical basis for Kant's moral imperative that I should always act in such a way that I would have all others act in the same way. Or, in the words of the Golden Rule, I should always do unto others as I would have them do unto me.

Given this basic, underlying moral law that follows from our quantum nature, there is a natural constraint on the fate of myself and the world that I help to create through my free decisions, and an objective criterion for deciding whether a given choice was a bad choice or a good choice. If it was a bad choice, it will lead eventually to a nonviable world, a world that cannot sustain an ordered coherence. Its values and its meanings break down and the moral equivalent of physical chaos sets in. To describe such a world, I might say something like "Everything is falling apart."

If it was a good choice, the world that it coauthored will be enriched; it will have a new ordered coherence, which I may articulate by saying something like "I've finally got my life together." But it is only when the results of my choice are in, when I can see and articulate the *meaning* of my choice (its meaning both for me and for all the others with whom I am in relationship), that it then becomes, historically, a good or a bad choice. Its goodness or badness comes into being when its results can be weighed (articulated) on the scale of successful or failed world creation.

If I choose to have the affair with my lover and in consequence my husband feels betrayed, our marriage breaks down, our children become disturbed, and I am so racked with guilt and despair that my world fragments, I will discover that I made a bad choice. The failed (fragmented) world to which my choice led will color its meaning.

If, on the other hand, the choice to have the affair has the very much more unlikely consequence that, through its having happened, both my husband and I reassess and revitalize our relationship and our marriage becomes stronger than before, I will discover that I made a good choice. It led to a successful world.

This line of argument does not, however, lead to the soft conclusion of moral relativism—"if it works, it must be good." Were there no human nature, and were it not a feature of that nature (our conscious nature) to be inextricably bound up with each other at the heart of our being, moral relativism might apply. But given that we are all "stitches in the same fabric," that in hurting others I hurt myself, there is an objective constraint on what makes a successful world.

Ultimately, no world will succeed in which the behavior of some hurts others, or in which the behavior of all hurts that larger process of increasing, ordered coherence in the universe at large, what Prigogine calls his "evolutionary paradigm."[18]

This constraint on what makes a successful world, and thus on what makes a choice good or bad, may not satisfy those who would have morality be a rigid structure of black-and-white dos and don'ts imposed from outside, a structure that says, for instance, that it is *always and ever* a bad choice for a married woman to take a lover. That kind of morality may be necessary for those who, for one reason or another, cannot live at the edges of their freedom, but it is not creative.

An evolving "quantum morality," or "creative morality," is necessarily more pluralistic than that. It not only allows, but incorporates as an important feature of what it is about, that there may be more than one way to respond in any given situation. In response to any moral challenge, there may be many choices that are, to some extent, "good" choices. It is an essential feature of our freedom, bound up with the purpose of our creativity, that we may try out any or all of these possibilities until we discover which is best of all, or the best so far possible.

Our free and undetermined moral choices and the worlds that they coauthor are like an electron's virtual transitions. They are experiments in reality creation. But we, unlike electrons, have a memory and can learn from our experience, so our experiments can have an aggregate effect. Some will succeed and go forward as lasting contributions to the next and better world; others will be lost to that process.

It is this capacity of the quantum self to pluck reality from multiple possibilities—the capacity to make experimental worlds, some of which will be improvements on the last, and our ability to articulate (through self-reflection) what made them so—that *essentially* links our freedom and our creativity.

The value of discovering the meanings that attach to my choices is that this discovery (this articulation) takes me back to the moment of freedom in which I made the first choice, the moment of decision that led to a chain of choices that became part of my life-style and of what I value—my world. In going back to that moment, I go back to the possibility of making some *other* choice, some other self and world. This is the reason why psychotherapy or any other thorough process of self-analysis can give us the possibility of "rebirth." *Rebirthing* means starting again, with the promise of a whole different life and world. It is related to the religious theme of renewal and rebirth—the Christian Easter, the Jewish Passover—which holds out the promise of new life.

It is our essential freedom, the fact that each choice we have made is only one of several possible choices we *might* have made, which allows this rebirthing and which gives each individual a crucial role to play in the gradual evolution of consciousness—the gradual increase of ordered relational holism as manifested in the worlds that we make.

Surely this is the meaning of our individuality, and a concept that gives a natural direction to the development of any human psychology that takes the creative role of human nature into account. As Charles Taylor expresses it:

> The prospect of psychoanalytic theory which could give an adequate account of the genesis of full human responsibility . . . with a truly plausible account of the shared subjectivity from which a mature cohesive self [and its world] must emerge, is a very exciting prospect indeed.[19]

CHAPTER 14

OURSELVES AND THE MATERIAL WORLD: QUANTUM AESTHETICS

> Perhaps the coming together of our insights about the world around us and the world inside us is a satisfying feature of the recent evolution in science. . . .
>
> —Ilya Prigogine
> *Order out of Chaos*

In many ways, the example of the creative dialogue between the child and his clay pot raises the whole large question of our relationship as conscious beings to matter, to the material world, and its relationship to us. To what extent does our conscious involvement influence the unfolding of material reality, and in turn, how much does the surrounding material world leave its mark on the unfolding of our own?

Could we, in short, be the human beings that we are without the material world's being as it is, and to what extent does our dialogue with the material world shape both matter and our humanity? The answers to these questions have important implications for our whole attitude towards the material environment.

At the simple, classical level of daily activity, we clearly have an effect on matter. We mold it and make it; we digest it, burn it, cool it, break it down, and build it up again. And at this same simple level, matter and its processes clearly have an effect on us.

The food that I eat can poison or nourish me; the rock lying on my path can trip me and break my toe; the locked door at the side of the house can stop me from getting into the garden. More basically still, the chemicals in the earth and the sunlight that shines upon them are necessary to the existence and functioning of my body, as are the oxygen that I breathe and the water that I drink. My body and all its physical needs are clear evidence of an essential dialogue between myself and the material world.

But the dialogue between the child and his pot, while retaining all these classical elements, has another dimension, a dimension that arises out of the child's *conscious* dealings with the material world. Through our material creations—our clay pots, our tools, our clothes, and our houses—we impute human meaning to the material; we bring it into our world of purposes and goals and thus transform it. But in doing so, we transform ourselves. Through making his clay pot, the child creatively discovers both that in the clay which could become this particular pot and that in himself (his sense of beauty and his skill) which could bring it into being.

Without the child's conscious intention, the clay would never have become a pot, but equally, without that pot, the child's sense of beauty would never have been incarnated. In some very important sense, child and pot gave birth to one another. The same is true, to a greater or lesser extent depending upon our involvement with them, of all the material artifacts in our environment. In making them, we make ourselves; through living with them we discover ourselves.

The most immediate aspect of the material environment with which we are in such a mutually creative dialogue is that of our own material bodies. The body has a whole complex range of needs, and it is through responding to those needs, through meeting them or thwarting them, that we not only satisfy or fail to satisfy their immediate requirements but also that we first discover ourselves and our world.[1]

When an infant first sucks at his mother's breast, he not only satisfies his immediate need for food but also lays the whole first foundations of his sense of self and others. He both becomes and *creatively* discovers that he is a person whose basic hunger can or cannot be met. He discovers, in Melanie Klein's terms, that the breast is a "good breast" or a "bad breast" and through that discovery becomes an infant whose world is a good world or a bad world. When he pushes on the breast and its surface yields to the pressure of his fingertips, he discovers and becomes a person who can influence the world. The same happens when he kicks against the side of his crib, but now he realizes that the world can hurt and that his power has limitations.

The breast and the crib are the world of the given. They are material things through which the infant discovers himself and his world, and through doing this he creatively discovers their *meaning* for him (their role in helping to integrate his perceptions), but they are not themselves of his own making. They are not his own creations. They can be meaningful for him because they meet or thwart his needs, but he has not evoked their objective, physical reality. That will come as he grows older and he begins to make the artifacts through which he can independently meet his own bodily needs—his tools, his methods of transport, his clothes, his shelter.

Each of the artifacts that we make in attempting to meet our direct bodily needs has an obvious functional purpose, and at this most primary level any object that is made can be judged a good or a bad object by whether it does or does not fulfill its function. A clay bowl that cannot hold soup is a bad bowl; a table that is too low for comfort or too unlevel to rest dishes upon is a bad table; a house that cannot keep out the rain is a bad house; and so on.

But because we are conscious beings with an equally strong set of needs for integrating our experience, for seeing ourselves reflected in our world, and for evolving towards ever greater ordered coherence in our world picture, the artifacts that we make must also fulfill a role of world creation. The clay bowl must somehow in its shape or texture reinforce and possibly develop our sense of beauty; the table must reinforce and develop our sense of balance and proportion and express our concepts of hospitality or communal eating; the house must express and nourish our sense of home.

These more recognizably "human" needs—what might be called the need for a life-style—follow directly from the physics of consciousness, from the fact that as the brain's quantum system maintains its dynamic ordered coherence, it naturally tries to draw everything that passes through it into its own integrated system.

Our artifacts, our reflections on those artifacts, and the habits we evolve around their use all get woven into the relational whole that is our world, just as the vitamins and minerals that we eat and the air that we breathe get drawn up into the dynamics of the relational whole that is the living body. And just as the body needs to change and grow in response to its environment (its evolutionary drive), so, too, consciousness needs to expand by forming ever greater relational wholes from the data of its world. The physics of consciousness and the physics of life are the same. Both are Fröhlich-style Prigogine systems.

When we judge an artifact's merit—when we say that a bowl is a good or a bad bowl, a table a good or a bad table, or a house a good or a bad house—we are really asking whether it meets both types of needs that led to its manufacture in the first place. These needs include the functional (whether it works) and the more "human" (whether it reflects our nature and enhances our world). The more human type of need might properly be called the aesthetic. It has to do with the "feel" of objects, the feelings they evoke in us, and with values, like beauty or even spirituality.

There are easy, quite mechanical, and obviously objective criteria for judging an artifact's function. No two people would disagree that a bowl that cannot hold soup is of little use. At the same time, the whole experience of modernist design in architecture and city planning, and of mass production in the manufacture of goods ranging from bowls and tables to clothes, motorcars, and houses, has exposed the inadequacy of mere function.

There is something ugly and brutal about the merely functional, about tables that are simply flat surfaces with legs and made of plastic, or about those ubiquitous concrete tower blocks that scar so many of our inner cities—what modernist architect Le Corbusier called his "machines for living." Their intentionally mechanical design excludes all consideration of any human factor in their use and reinforces the sense of alienation that has its roots in the whole mechanistic bent of our Newtonian culture. Artifacts that contain nothing of the human (nothing of human consciousness) reflect nothing back when we deal with them. They are not in dialogue with us and they cannot meet our need for creative self-discovery.

But the criteria for judging whether a bowl or a table or a house meets our aesthetic needs seem at first glance more elusive. If they have been made spontaneously by ourselves, they will, by the very nature of that creative dialogue between ourselves and matter when we shape something, naturally express (and create) ourselves and our world. But if they have been made from design or made for us by others, they may not. They may then be "insensitive" to our needs, or even stifle them. How can we judge?

There have been various aesthetic philosophies through the ages that have tried to answer this question. Plato believed that a thing was beautiful if it reflected its original in the Realm of Forms, his equivalent of a cosmic blueprint for every existing thing in this world. Aristotle took the beautiful to be that which strove towards a "golden mean"—his principle of nothing in excess—which applied to both art and morality and which was bound up with his general view that there was a direction and a purpose (a teleology) in the natural unfolding of things.

For the Romans, the beautiful was that which reflected the underlying principle of their law. This law had as its goal achieving the greatest possible internal coherence of society, and all their principles of efficiency derived from this end rather than from mere function.[2] Similarly the great Christian architecture, the cathedrals and the Gothic arches and the towering spires, had as its guiding vision the love of the Virgin or the idea of God on high whose love and wisdom directed all things below.[3] In England, the great Victorian buildings expressed the power and extent of the empire.

We have no such visions today. The laws of Nature as seen through the Newtonian perspective strive towards nothing. They simply *are,* cold, mechanical, and determinate. The cosmology of the Church, which gave us God on high, has been discredited by Galileo and Copernicus, and Plato's Realm of Forms dismissed as so much Greek mysticism. And even the humbler vision of man's place as a creature of Nature has been undermined by the move towards urban life and mass production. Few of us have any contact with the origins of the food that we eat or the clothes that we wear, and we are often ignorant of the natural processes that produced them (in those increasingly rare cases where they are themselves natural products).

In consequence, many people in the modern era have argued that there are no objective standards for judging aesthetic merit, no clear way to say that one bowl is more beautiful than another or one house more satisfying to live in than another. Such things are said to be a matter of taste, and "there's no accounting for people's taste." Instead, it is better that we concentrate on the mechanical and the functional, where there are clear standards.

But each of us as a conscious being does carry the natural within himself, whatever his life-style or circumstances. We carry it within the physics of our consciousness, which is the same physics as that of life itself. If we relate the aesthetic dimension of the objects we use in our daily lives to the aesthetic needs of this consciousness, then we may find again within its physics certain natural criteria for a value like "beauty," just as there were found to be natural criteria for what is "good" in the realm of the moral. It is important to remind ourselves, though, that there may be many quite different expressions of beauty that meet these criteria, just as there are many possible forms of behavior that might meet the criteria for "good" behavior.

A "quantum aesthetics" would allow necessarily for the possibility of many equally valid aesthetic styles, though we may find that there is a "constraint of the natural" that underpins all our aesthetic needs and gives us an objective foundation for judging whether a given style or object meets them. If an aesthetic style expresses and cultivates the natural in us—the nature of our consciousness—it succeeds; if not, it fails. We can look for the relevant criteria for judging this by reminding ourselves of the basic features of Fröhlich-style Prigogine systems, of the ordered, coherent quantum system that is the physical basis of consciousness.

The most important *dynamic* feature of any Prigogine "open system," quantum or otherwise, is that it is poised delicately at a very critical dividing line between the static and the chaotic. This is what Prigogine refers to as "far from equilibrium conditions."[4] If it had less energy going through it, it would run down and the matter caught up within it would become inert, without order or meaning. If it had more energy, it would veer off into too much turbulence and become mere "noise." A whirlpool that runs down ceases to be a distinct, organized pattern in the water and merges homogeneously with its unstructured surroundings; a whirlpool that becomes too excited dissipates into chaotic turbulence—it "comes apart at the seams."

This same fine balance is maintained in any conscious system, and in psychological terms represents the dividing line between boredom on the one hand and confusion on the other.

An artifact that bores us does so because there is insufficient "movement" either in its design or in the materials of which it is made. This is very obvious in paintings that have no "life," but it can also be true of houses or bowls. A functionalist building designed according to strict mechanical principles will have no eccentric edges. All the corners will be square, all the ceilings of just the right height to clear the head of the tallest person, all the windows and doors absolutely symmetrical.

There will be no surprises, nothing to catch the eye and thus to stimulate the mind.

This boredom is all the more pronounced if in addition to having mechanical lines the building is made of a synthetic material, such as concrete, which is homogeneous throughout. Living systems and conscious systems are not homogeneous. With the rare exception of clones, no two living systems are the same, and this is all the more true of conscious systems, which have been differentiated in dialogue with their experience.

So a functionalist building made of concrete in no way reflects either the dynamics or the texture of consciousness. Its consequent boredom factor can be seen as objectively bad (ugly) because it violates the constraint of the natural. This violation can result in actual pain for the conscious beholder, and it is just this sort of pain that we often see expressed in the violence and vandalism directed towards such buildings.

The same is true of those tedious housing estates or "new towns" where row upon row of identical units are crowded into small spaces by designers who try to apply the techniques of mass production, and its assumption that we are all the same, to the conditions in which we live. There is insufficient complexity in such structures to reflect the complexity and the eccentricity (individual differences) inherent within the physics of consciousness. No two living Bose-Einstein condensates are the same, and we have a need to see this individuality expressed in our surroundings.

In these cases, the pain resulting from boredom is compounded by the physical effects of crowding. When an electron is confined in too small a space, it goes "berserk." Its position becomes too fixed, and thus the Uncertainty Principle (we can't know both a particle's position and its momentum) requires that its momentum become too great. Being quantum systems, conscious systems suffer similarly from too much crowding.

Equally, a building or a painting that has too little structure is confusing and can evoke feelings of pain. The Dada movement in modern art, which fought all structure and convention in an attempt to express the irrational world of the Freudian id, had this quality and most of its products could be judged objectively ugly. A novel like James Joyce's *Finnegans Wake,* which offers too many associations of ideas per page, is simply very confusing.

Buildings constructed haphazardly with little thought for design criteria are rare, if for no other reason than that they would be likely to fall down. Many inner-city neighborhoods, however, have a cluttered, chaotic look because of ill-considered mixtures of housing styles and textures, and many cities themselves have simply spread without rhyme or reason. In America, where few cities have any green-belt planning restrictions, this clutter is allowed to spread indiscriminately into the surrounding countryside. People see little of their own nature in such confusion and the sense of alienation in these neighborhoods and cities runs deep.

Similarly, a house packed with too much furniture, too much unintegrated decoration, too many incompatible colors, or simply too much clutter ceases to feel like a home. When my children's playroom becomes too disorganized, with none of the toys in the right place, they complain that "it's too messy in there" and gravitate towards the living room. The playroom's complete lack of order violates their natural, conscious need for the right level of ordered coherence in their surroundings.

A more positive example of a design concept that reflects the dynamic nature of consciousness might be the Japanese garden, where the use of flowing water (waterfalls) and the cascading effect of small hills and rocks poised on many levels suggest a sense of balanced movement throughout. The delicate and yet forceful superposition of colors and textures among the chosen trees and plants has the same effect, and one feels both still and stimulated in such surroundings.

For the Japanese, the garden is a quasi-religious artifact, a material expression of the human spirit in tune with Nature, and it is little accident that it incarnates the qualities of human consciousness. It does so through the use of entirely natural materials, but there are structures in the West that create the same effect with more artificial ones. I think for example of the marvelous early glass buildings of Mies Van Der Rohe, buildings that appear to have been *sculpted* out of their material substrate rather than simply assembled.

This same sculpting effect is achieved in many of our more humble daily artifacts through prolonged use, provided that they are made from materials that slowly give way in dialogue with the human form or human activity.

A hammer newly arrived from the factory has little "character," but if it has a wooden handle it will eventually take on the shape of its renewed contact with the laborer's hand and with his labor. A new leather shoe may have considerable beauty if well designed, and especially if handmade, but it fully comes into its own as an aesthetic object only after being worn and slowly sculpted to the wearer's gait and his life-style. It is then that it becomes a fitting subject for a van Gogh painting or something that we might even consider hanging above the dining room table.

This "worlding," as Heidegger would call it in his aesthetic theory,[5] of the hammer or the shoe, the Japanese garden or the Mies Van Der Rohe building, the way in which an object comes to express—to shape and take the shape of—the world of its users, reflects more than just the way conscious systems are poised between the static and the chaotic, the boring and the confusing. It also reflects the multiple potential (the hidden "depth") of the quantum superposition effects that are an essential feature of any quantum system.

Quantum systems are like poems, always pregnant with as-yet-unrealized meanings, always begging evocation and interpretation. Schrödinger's cat is both alive *and* dead, and it is that duality which makes him such an engaging character. My thoughts may now take this form and now that, make an association here and a different one there, and it is this free, shifting quality that makes us so creative. Similarly, the world that I create around myself and the objects in it, if left to evolve naturally, will express this poetic quality, and those that are designed for me will feel "beautiful" only if they reflect it.

The mechanical world has none of this free and indeterminate depth. It is fixed and "shallow," without hidden perspective. Harsh, functionalist architecture or purpose-built, identical housing units are like the crudest prose. We can see and feel all there is to them in one cursory glance, after which there is nothing with which to engage in a mutually creative dialogue. Similarly with plastic bowls and furniture, plastic table tops and toys.

Plastic is a fixed and one-dimensional substance, and things made of it are uniform throughout. It takes on any determined shape we may initially wish to give it and then it has nothing of itself to give back. It does get scratched and dirty, but it doesn't wear naturally or mold itself to the changing shapes of bottoms or the incessant pressure of fingers. It doesn't become worn where the elbows have rested upon it, or slightly wobbly with the potter's moods, or altered where the child has held it within his imagination and his grasp through endless hours of play. It has no more capacity for dialogue and world creation than its stronger counterpart, concrete.

World creation itself, in this case the mutually creative dialogue between persons and their material implements and surroundings, is bound up with the very most basic feature of the physics of consciousness, its ordered coherence and its deep "instinct" to increase and expand that quality. Material things that cannot be integrated into our world because they are too boring or too confusing, too rigid or too anonymous, do nothing to stimulate consciousness or to increase its bounds. One crucial consequence of this is our reduced ability as conscious persons to meet each other and to share a world through our material creations and their use.

Heidegger has portrayed vividly how through van Gogh's painting of the peasant's shoes we enter the world of the peasant and his labor, how we share his relation to the earth and sky, to other peasants, and to the German folk tradition.[6] The same would be true to a lesser extent if we just encountered the shoes, or the peasant's plow, or the chain with which he controlled the bit in his horse's mouth. This is why we are drawn to antiques and rusting old artifacts. We are drawn to the worlds of those who used them.

Equally, when we visit the Colosseum or Pompeii, we enter the world of Rome and her highly integrated civilization and of the Romans who lived and died in these places. We can feel the glorious rush of the chariots and the terror of the Christian slaves, the daily laughter and lovemaking of the Pompeians and the dread as molten lava swept down the mountainside.

If we enter Chartres or Notre Dame or Saint-Michel, we encounter all the power and the glory and the majesty of Christianity at its height and we can feel the presence of aristocrats and peasants who for centuries prayed, lit candles, and offered up their deepest yearnings among these arches and incense-impregnated walls. One simple, worn, leather prayer cushion tells the stories of several hundred thousand lives.

When, one day, future generations excavate our concrete apartment towers and their tin-and-plastic debris, their polyester curtains, and their wrinkle-free wash-and-wear shirts, what will they be able to find of the people who lived there? What of all the quietly desperate people who inhabit the rows of identical, purpose-built units on identical, characterless streets? What do these artifacts reflect of their owners' lives and loves, their labors and their visions? As Rilke expressed it over seventy years ago, we are slowly losing "the honey of the visible":

> Even for our grandfathers a house, a fountain, a familiar tower, their very clothes, their coat, was more, infinitely more intimate; almost every object a vessel in which they found something human, or added their morsel of humanity. Now, from America, empty indifferent things crowd over to us, counterfeit things, the veriest dummies. A house in the American sense, an American apple, or one of the vines of that country has *nothing* in common with the house, the fruit, the grape into which we have entered the hope and meditation of our forefathers. The lived and living things, the things that share our thought, these are on the decline and can no more be replaced. We are perhaps the last to have known such things.[7]

Rilke was wrong to think such spiritual desecration only an American problem, or perhaps even American in origin. It need not be that we are the last to have had a lived dialogue with the material world. But his words do express a pain that many of us feel in the face of the ever-encroaching ugliness and anonymity of the brutal, the plastic, and the boring in our material surroundings.

This poverty of the material isolates us not only from other cultures and generations, which can find little of us in our artifacts, but also from each other, now, in daily, simple ways.

When I speak of these things, of the material things that do reflect the physics of consciousness and those that don't, of the poverty of the latter and the alienation that follows in their wake, I am reminded of two local parks that I knew well when I lived in inner London. One park was the responsibility of the local city council, the other of the residents of nine local streets.

The city park had been designed and planted haphazardly over the course of a century, and was maintained by salaried council gardeners who lived outside the area. It was a large, flat space with a curious mixture of bushes and trees selected with little apparent thought, and an unshaded play area paved in asphalt and surrounded by a high chain-link fence. Nothing had been done to separate the park visually from the unsightly back views of high brick and concrete buildings on the adjacent road.

This park was boring: It had neither movement nor coherence in either its design or its planting scheme, and it had been maintained over the years by professional workers who put nothing of themselves into it. It expressed no creative dialogue with its surroundings and evoked none from those who frequented it. It was not a part of anyone's world. It was the sight of much personal violence—drunken fights, muggings, and riots—and it was strewn with litter and very frequently vandalized.

The smaller neighborhood park had been designed, built, and planted entirely by local residents and their children after receiving a hard-won grant from the city council to convert what had been a derelict building site. Against the opposition of council planners who dismissed their ideas as "silly frills," the residents imported truckloads of earth to create a hill on part of the site. They chose plants whose colors and textures blended and set each other off; they added brightly colored play equipment and surrounded it with benches for mothers to sit on and trees to shade them from the sun. The entire park was surrounded with a Victorian-style wrought-iron railing, which the residents painted themselves.

For two years the new shrubs were watered by local people in rotation, and other teams kept the weeds away. But slowly most original founders of the park moved out of the neighborhood and those few left behind feared that council predictions of eventual vandalism and misuse would come true. It was a rough, working-class, and largely immigrant area, where few people who used the park knew its local origins. But the fears were unfounded. The same children who vandalized the city park treated the smaller park with care. Immigrant mothers who showed little regard for litter baskets on the main shopping street kept a sharp eye out for rubbish in the residents' park. Mysterious "someones" kept the weeds down and the park has remained beautiful.

The small park was a world. It originated in a neighborhood's concern for its immediate environment and grew out of a creative dialogue between local residents and a derelict piece of land. Through that dialogue, both a wider neighborhood and a park were created. Residents who were no part of the original scheme were drawn into the park's ordered coherence and shared a world with those who first built it. The park remains, like the Colosseum or Chartres but on a humbler scale, a "lived and living thing."

The residents' park was originally the idea of a small band of sensitive, educated people, many of whom were active in the London theater. They were scorned as an elite by council officers who thought they knew better what that kind of neighborhood required. Their first choice was a parking lot, their second a small-scale version of the bleak city park. They felt that the Irish, Indians, and Pakistanis of inner London had no need of Victorian railings and trees with odd Latin names.

City planners, administrative officials, and government architects dealing with inner cities throughout the world exhibit the same arrogance. They might themselves live in imaginative houses in well-planned neighborhoods, but conclude nonetheless that concrete apartment towers and flat open spaces are all "the masses" can appreciate. The same kind of arrogance, or possibly just thoughtless sloth, results in badly designed schools and recreation centers, in inner-city public eating places that offer nothing but greasy and tasteless "plastic" food, in inner-city clothing outlets that sell badly cut, synthetic clothes. "Those people wouldn't know any better."

But even the masses carry Nature within the physics of their consciousness, and have a need to see that Nature reflected in their environment. In past, more rural, and simpler times, these same people would have made their own clothes and artifacts. Their products would have filled their own homes and enriched their own lives. Today such creations are known self-consciously as "folk art" or "peasant crafts," and their possession is a privilege of the better-off.

Today's urban workers and "peasants" are dependent upon others to supply their needs, upon others to design their clothes and chose their food and build their houses. But it is *their* world that is impoverished when all they are offered are plastic and concrete things in which they cannot discover themselves, and their alienation that eats away at the fabric of modern society.

CHAPTER 15

THE QUANTUM VACUUM AND THE GOD WITHIN

We live our lives inscrutably included within the streaming mutual life of the universe.

—MARTIN BUBER
I and Thou

As a very young child, when I looked up at the night sky I saw Castor and Pollux, Orion and Cassiopeia. They were not mere star configurations, but people, heroes and heroines about whom I had read exciting stories and whose brave deeds inspired so many of my games and fantasies. When the wind blew or a storm struck, it was Poseidon venting his anger or Zeus having a tantrum.

With a different inner eye I searched for the Christian heaven and wondered on which of the many bright lights in the sky God actually had His throne. He, too, blew up winds and made storms when he was cross with the world, but it was His world, He had made it, and I trusted Him to take care of both it and me.

Now I lay me down to sleep;
I pray the Lord my soul to keep.
If I should die before I wake,
I pray the Lord my soul to take.

My mother was a classics teacher and my grandfather a devout Christian, and between them they peopled my childhood with gods and faith. But as I grew older and learned more about the way the world "really is," about astronomy and cosmology and evolution, the faith of my childhood (and the whole world constructed through it) came to seem like so many fanciful stories. The night sky became a thing of cold indifference and my own existence a matter of accident and insignificance so far as the world was concerned. In this, my own experience mirrored that of my generation—to a large extent that of the past few generations. Our science was at cross-purposes with our traditional faith.

Many people have argued that the discoveries of modern science need have no impact on traditional religious faith. As the British physicist Brian Pippard puts it, "The true believer . . . need not fear—his citadel is impregnable to scientific assault because it occupies territory which is closed to science."[1] On this view, faith and reason represent two different worlds, speak different languages, and maintain different notions of truth. Each is alien to the other and neither can learn from or refute the other. But an ostrichlike "I don't want to know about it" approach to science is true neither to the history of religion nor to the personal experience of most people.

Nearly all the great religions have embraced and reflected the "science" of their day, or at least the current understanding of Nature and her forces together with the most current knowledge of human nature and psychology. This is because one central motivating force behind any religious perception is the attempt to form a coherent picture of the world and one's own place within it.

Thus the ancient Greeks, who were obsessed with natural forces—winds, storms, earthquakes—and with human impotence in the face of them (fate), conceived of their gods and goddesses as superior humanlike embodiments of these forces and of themselves as the playthings of the gods. Like themselves, the gods were brave and cunning, sometimes peevish and vindictive, and the challenge was to win them over or to outsmart them.

The Buddhist "science" was a science of consciousness, a science of states of awareness. It was concerned with how to see through their illusions, how to control them, and the Buddhists thus conceived of the universe as something like the all-embracing ground state of consciousness, a consciousness from which human consciousness had become split off. The challenge was to return to the ground state, to achieve union with it and thus to achieve nirvana—timelessness and awareness/unawareness.

The Christian tradition, like the Jewish one on which it was founded, was concerned with social unification and order—the Law, or oneness in the body of Christ. The foundations of this tradition, however, embraced the earth-centered cosmology of Ptolemy and the Platonic conviction of a split between this earthly world of matter and the world of spirit. For the world of matter, the Church Fathers happily adopted the major features of Greek science, but they rejected Aristotle's notion of a universe that had always existed because it clashed with the biblical creation story. To some extent they similarly rejected Aristotelian teleology—the notion that matter had a sense of purpose or direction (Aristotle's "final cause")—because it clashed with the split between matter and spirit.

For the world of spirit, Christianity conceived of a transcendent God who let His influence be felt through the forces of the heavenly spheres inhabited by various orders of angels (the basis of astrology) and through the earthly intervention of His Son. This transcendent God was Himself outside time and history. No laws of physics constrained His imagination. His Son took on a material form, but He, too, was outside the laws of physics and His kingdom was not of this world. Hence the virgin birth, the miracles, and the bodily resurrection.

Until the seventeenth century, there was little distinction between priests and scholars, and little scientific challenge to the physics or the cosmology of the Church. But with the explosion of modern science, this challenge became impossible to ignore.

Slowly, the story of the creation, the notion of human uniqueness, the idea of an earth-centered universe and hence one to which a transcendent God would devote any special attention, the credibility of the miracles, and that of the bodily resurrection have all become problematical. The Church has clung tenaciously to what some of her more modern priests call "a Sunday school religion," but large numbers of her followers have been assailed by doubt.

It is no longer possible to believe in *both* the discoveries of modern science and in the traditional dictates of the Church, and, for increasing numbers of people today, science and psychology have taken the place of traditional religion. We want, perhaps more than ever, to understand ourselves and our world, to know the history of the universe and of our place within it, to form a coherent picture of how we should behave and towards what goals we should strive. We long to know what is valuable and what is not. But we look increasingly to science to tell us these things. When it has no answers, we feel lost.

Neither the mechanical physics of Newton nor the biology of Darwin has said much that might contribute towards a coherent picture of ourselves within the universe. Newton's physics has nothing whatever to say about consciousness or about the purpose or goals of conscious creatures. The mechanical world view did a great deal to undermine the certainties of Christianity, but had little of spiritual value to offer in their place.

Similarly Darwinian biology, whether in its original brutish and determinist form (survival of the fittest) or in its neo-Darwinian emphasis upon random evolution, has little to say about why we are here or how we relate to the unfolding of material reality, never mind about the purpose and meaning of any evolution of consciousness beyond the simple, utilitarian conclusion that consciousness seems to confer "some evolutionary advantage."

Mechanical science has given us a great deal of knowledge but no context within which we might interpret it or relate it to ourselves and our concerns. In the same way, technology has given us a much higher standard of living, but no sense of what living is about—no greater quality of life. Technology, like pure mechanical science, is value free; it is there for any and every use. In many ways that has been its strength, just as cold objectivity was the strength of Newton's physics—it separated the purposive from the mechanical and made it possible to see clearly what made the mechanical function. But this kind of science and technology tells us nothing about ourselves, and leaves us feeling alienated from our material surroundings. On its own, with no spiritual complement, it leaves us feeling alienated from each other and from our world.

It has been the argument of this book that quantum physics, allied to a quantum mechanical model of consciousness, gives us an entirely different perspective. This is a perspective from which we can see ourselves and our purposes fully as part of the universe and from which we might come to understand the *meaning* of human existence—to understand why we conscious human beings are here in this material universe at all.

If such a perspective were fully achieved, it would not replace all the vast poetic and mythological imagery, the spiritual and moral dimensions of religion, but it would provide us with the physical basis for a coherent world picture—and one that includes ourselves.

I said very early on, when discussing the problem of Schrödinger's cat, that quantum physics raises the question of consciousness and does so in a way such that consciousness becomes an issue for physics itself. Something about our conscious participation in the design of laboratory experiments evokes a given aspect of many-possibilitied quantum reality and causes that aspect to be real-ized, just as the child's conscious participation in the making of his clay pot evokes a particular pot (and a particular child).

But how far back, into what deep levels of mutual reality formation, does this creative dialogue between consciousness and matter extend, and how can we relate it to the physics of consciousness? To what extent can we see consciousness as having had a role to play in the making of objective, material reality—things that we can bump into, look at, and measure? On what level can we say that that same reality has had a creative role to play in developing consciousness?

In trying to answer these questions, it is necessary to clarify what we mean by consciousness.

In our human terms, the word *consciousness* is used to embrace a whole panoply of meanings and associations—mind, intelligence, reason, purpose, intention, awareness, the exercise of free will, and so on. Some of these uses clearly can be extended to describe the conscious behavior of higher animals and a few perhaps even to that of simple creatures like the amoeba. But when consciousness in this broad, fully human sense is used to describe the activity of either a transcendent or an immanent agency working to create or shape the material world from the beginning of time, it borders on traditional mysticism or theology. That is not the sense in which I am using it here.

Human consciousness in its fullest and broadest sense no doubt developed in a long evolutionary process from much simpler, very much more elementary forms of consciousness. If we want to understand the nature and dynamics of our own complex mentality, and its place in the wider scheme of things, we need to see its roots in these simpler forms and their dialogue with the material world. In tracing that heritage, we may gain some perspective on the whole history of which we are a part.

I have argued that at any level that we can recognize as existing in ourselves, the physical basis of consciousness rests on a very special sort of dynamic relational holism—a Fröhlich-style Bose-Einstein condensate in the brain, a coherent ordering of some bosons (photons or virtual photons) present in neural tissue or neuron cell walls. This quantum coherence makes possible the orchestrated firing of some or all of the 10^{11} neurons in the human brain and the integration of information to which their firing gives rise—thus giving us the unity of consciousness and, ultimately, the sense of self and world.

Without the ordered, Bose-Einstein orchestration of photons (or other bosons), there would be no sense of self and world; but equally, without the material components of the neural tissue, there would be no Bose-Einstein condensate. The two, quantum coherence (the ground state of consciousness) and neural tissue (matter), in relationship to each other, give the brain its conscious functioning capacity. This capacity is linked to all the neural networks that process data from the environment.

So at the level of consciousness found in ourselves and higher animals, the creative dialogue between matter and consciousness is obvious and crucial—neither is reducible to the other, and yet neither could function without the other.

Equally, and at a more basic level, this same ordered quantum coherence is thought to be present in all biological tissue, right down to the level of DNA itself. As we have seen, it is linked inseparably to the essential creativity of life. That creativity springs from the self-organizing capacity of all living systems (Fröhlich-style Prigogine systems) to take unstructured, inert, or chaotic matter from the surrounding environment and draw it into a dynamic, mutually creative dialogue that results in both more complex structure and greater ordered coherence. The coherence of living systems, then, both evokes a potential hitherto unrealized in the matter that becomes organized through it, and more fully realizes itself.

The ordered quantum coherence which is life hasn't the capacity for self-consciousness that we associate with the quantum coherence linked to higher brain functions. It would be an anthropomorphic projection to speak of its having a sense of purpose. But as Ilya Prigogine has argued, it does have a sense of direction—what he calls his "evolutionary paradigm."

Life seems always to create more life, more and greater ordered quantum coherence. And this is a clear antecedent of the purposiveness that we find in conscious systems like our own. It has the same physics, and through this physics we can trace our consciousness back to something that we share, in some very primitive sense, with any living thing. And at each level where there is ordered quantum coherence, there is a creative give-and-take between that coherence and its material surroundings.

So we conscious human beings clearly share some of our nature with all other conscious creatures. At a simpler level we share the basic physics of our consciousness with all other living things—we all have in common a mutually creative dialogue with the material environment. But the interesting question is whether life itself has any antecedent. Is the living world just a random offshoot of brute universal processes that are themselves wholly alien to life, or is there some early ancestor of the physics that becomes the physics of life? Can we trace our conscious ancestry back into the nonliving world?

I have argued earlier (in Chapter 7) that, ultimately, we can trace our consciousness back to its roots in the special kind of relationship that exists wherever two bosons meet, to their propensity to bind together, to overlap, and to share an identity. It is this propensity that makes possible the much more coherent ordering of more complex quantum systems (those found in life and human consciousness)—where millions of bosons overlap and share an identity, behaving as one large boson—but in its primitive form it is there whenever two bosons meet. Physicists working with photons call it the photon bunching effect, noting that photons emitted from any ordinary, noncoherent photon source arrive at a detector in bunches (Figure 15.1).[2] It is their nature to "socialize."

The bunching effect has moved German physicist Fritz Popp to

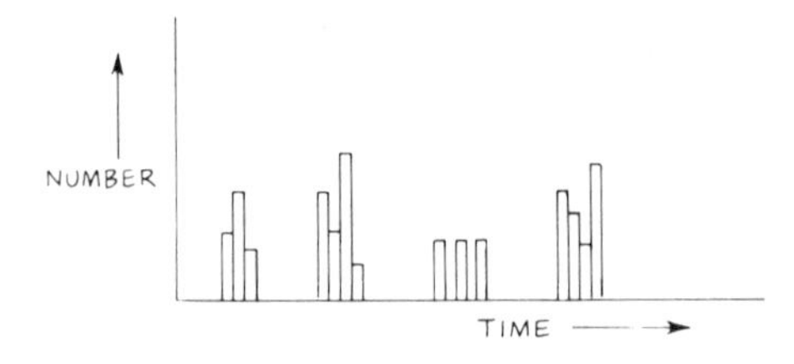

Figure 15.1. Photon bunching. If we have a chaotic (noncoherent) photon source, the photons arrive in clusters.

conclude that "the difference between a living system and a non-living system is the radical increase (an order of magnitude 20 times greater) in the occupation number of the electronic levels."[3] That is, in living systems photons are *very* much more (exponentially more) bunched together, literally "squashed" into a coherent Bose-Einstein condensate—whereas in the nonliving, they are less tightly packed. But this difference is one of degree, not of principle.

In the process of photon bunching we see the primitive antecedent of the coherence that becomes life, but on its own it is timeless—it has no sense of direction. That direction arises through processes described in the physics of "self-organizing open systems," Ilya Prigogine's physics—through the fact that within open systems, unlike those driven by entropy, order always *increases.* Living systems are such open systems, but their physics extends further back, back into the world of the nonliving.

A Prigogine open system, like a waterfall, needs to be driven by a flow of matter or energy running through it. It could not maintain its drive towards increased order in a static or homogeneous universe, a universe at equilibrium. Remember that creativity happens at far-from-equilibrium conditions.

But our universe is neither homogeneous nor static. One need only look up into the night sky to see clumps of galaxies and clusters of stars, all of which have vast reserves of gravitational energy, energy that can drive Prigogine self-organizing systems. As Prigogine's colleague Gregoire Nicolis says, "Gravity may therefore be regarded as a basic organizing factor in the universe, mediating the passage from equilibrium to nonequilibrium."[4] Gravity itself is a boson force field.

Thus bosons (gravitons) are a large-scale driving force that moves the universe towards greater order. At a more basic level still, they may even be responsible for the collapse of the quantum wave function, the problem highlighted by the enigma of Schrödinger's cat.

It appears, according to very recent work, that the wave function may collapse whenever two bosons overlap and share an identity (or when they cease to do so).[5] In this strict and limited sense, where the roots of consciousness can be traced back to wherever two bosons meet, it may after all be accurate to say that consciousness collapses the wave function. And this collapse is the very *most* basic of Nature's irreversible processes. This would be another, and still more primary, way in which bosons (the building blocks of consciousness) introduce a sense of direction into physics (Aristotle's teleology) right from the start, a sense of direction necessarily allied to the material world.

Bosons are, essentially, "particles of relationship." They are the fundamental building blocks of all Nature's forces—the strong and weak nuclear, the electromagnetic, and the gravitational. They are the most primary antecedents of consciousness, but they also bind together the material world.

The fundamental building blocks of the material world itself are fermions (for instance, electrons and protons), those "antisocial" particles that prefer to keep to themselves. Without bosons, fermions would seldom get together and build anything;* without fermions, bosons would have nothing to draw into relationship and thus nothing with which to order and structure their own more complex coherence. From the very beginning, then, from the most primary level of what later becomes the material world and the world of consciousness, the building blocks of matter (fermions) and the building blocks of consciousness (bosons) are necessarily involved in a mutually creative dialogue.

That which, in a far more complex form, later becomes us, is part and parcel of the basic dynamic through which the universe unfolds. With this understanding of the origins of consciousness—that it begins wherever two bosons meet—it may not be too wild to speculate that a gradual evolution of consciousness is the driving force behind that unfolding. This is not quite as strong as saying that Mind created the world, but it is suggesting that the elementary building blocks of Mind (bosons) were there from the beginning, and were necessary partners in that creation. In creating themselves (fulfilling their nature as "relationship"), they evoke the world.

This proposed "genealogy of consciousness," which traces the roots of our own complex mental life back to their origin in simple boson relationships and the origins of our universe back to the mutually creative dialogue between bosons and fermions, lends a new kind of physical interpretation to one version of what cosmologists call the Anthropic Principle. There are many versions being proposed, from the "weak," which says simply that the universe has to look like it does to us because it is we who are looking at it, to the "strong," which argues

---

*There are exceptions, such as when two fermions overlap and behave like a boson—this happens in the electron rings of covalent chemical bonds.

more religiously that some form of intelligent life like ourselves *had* to result from the unfolding universe.[6]

The genealogy of consciousness that I am proposing supports, in a limited sense, John Archibald Wheeler's version, which is called the Participatory Anthropic Principle. This principle argues that "observers are necessary to bring the world into being."[7]

In my terms, the "observers" are not only full-fledged, intelligent, conscious beings like ourselves, but also all of our long line of predecessors, going right back to simple boson pairs. Without those bosons there literally could not be any universe such as we know it—they are the glue that holds things together. But without complex creatures like ourselves, the universe might unfold less extensively, or at least much more slowly. As Ilya Prigogine says, "It is interesting that with an increase of complexity, going from the stone to human societies, the role of the arrow of time, of evolutionary rhythms, increases."[8]

This capacity to increase the rhythms of evolution, specifically of evolving consciousness, may suggest a reason for human existence. It may hold out the key to *why* we are in the universe, and give us some good notion of exactly *where* we fit into the general scheme of things. To understand this fully, we need to see the link between the physics of human consciousness that I have proposed in this book and the physics of the quantum "vacuum" as proposed by quantum field theory.

The quantum vacuum is very inappropriately named because it is not empty. Rather, it is the basic, fundamental, and underlying reality of which everything in this universe—including ourselves—is an expression. As British physicist Tony Hey and his colleague Patrick Walters express it, "Instead of a place where nothing happens, the 'empty' box should now be regarded as a bubbling 'soup' of virtual particle/antiparticle pairs."[9] Or, in the words of American physicist David Finkelstein, "A general theory of the vacuum is thus a theory of everything."[10]

After the Big Bang in which our present universe was born, there was space, time, and the vacuum. The vacuum itself can be conceived as a "field of fields" or, more poetically, as a sea of potential. It contains no particles, and yet all particles come about as excitations (energy fluctuations) within it. By analogy, if we lived in a world of sound, the vacuum could be conceived as a drum skin and the sounds it makes as vibrations of that skin. The vacuum is the *substrate* of all that is.

The exciting realization, from the point of view of understanding consciousness, its roots, and its purpose, is that one of the fields within the vacuum is thought to be a coherent Bose-Einstein condensate, that is, a condensate with the same physics as the ground state of human consciousness.[11] Further, excitations (fluctuations) of this coherent vacuum condensate appear to have the same mathematics as the excitations of our own Fröhlich-style Bose-Einstein condensate.[12]

Understanding this might well lead us to conclude that the physics which gives us human consciousness is one of the basic potentialities within the quantum vacuum, the fundament of all reality. It might even give us some grounds to speculate that the vacuum itself (and hence the universe) is "conscious"—that is, that it is poised towards a basic sense of direction, towards a further and greater ordered coherence. If we were looking for something that we could conceive of as God within the universe of the new physics, this ground-state coherent quantum vacuum might be a good place to start.

For some people the idea of a transcendent God who creates, and possibly controls, the universe from a vantage point outside the laws of physics, from beyond space and time, will always remain appealing, and there is nothing to stop them from speculating that this God preceded—and possibly caused—the Big Bang. This is a perfectly tenable position, though it leaves us with a God who Himself undergoes no creative transformation, who is not in dialogue with His world. Such a belief must remain wholly a matter of faith. There is no way, from our post–Big Bang position, that we can know who or what preceded it.

But if we think of God as something embodied within, or something that uses, the laws of physics, then the relationship between the vacuum and the existing universe suggests a God who might be identified with the basic sense of direction in the unfolding universe—even, perhaps, with an evolving consciousness within the universe. The existence of such an "immanent God" would not preclude that of a transcendent God as well, but given our knowledge of the universe the immanent God (or immanent aspect of God) is more accessible to us.

This immanent God would be at every moment involved in a mutually creative dialogue with His world, knowing Himself only as He knows His world. It is the concept of God proposed most strongly in this century by Teilhard de Chardin,[13] and more recently by "process theology."[14] In terms of this concept it makes sense to speak of human beings—with our physics of consciousness, which mirrors the physics of the coherent vacuum—as conceived in the image of God, or as partners in God's creation. As Teilhard puts it:

> We are not only concerned with thought as participating in evolution as an anomaly or as an epiphenomenon; but evolution as so reducible to and identifiable with a progress towards thought that the movement of our souls expresses and measures the very stages of evolution itself. Man discovers that *he is nothing else than evolution become conscious of itself,* to borrow Julian Huxley's concise expression.[15]

Like the ground state of human consciousness, which is coherent but in itself "uninteresting"—without features—the coherent quantum vacuum contains within itself all potentiality.* It can realize that potentiality, however, only through the fluctuations within itself, excitations that lead to the birth of particles and their relationships. In ourselves, these excitations give birth to thoughts. Our thoughts are the "interesting" and creative aspect of consciousness, but they purchase these qualities at the expense of splitting off from the coherent ground state. Similarly with particles.

In religious terms, this splitting off might be equated with alienation, or the Fall. Such a Fall is the prerequisite of all creation (or knowledge), but it means leaving the Eden of total fusion.

We have seen, however, that the basic evolutionary drive of the universe, or at least of that aspect of the universe that results, eventually, in living systems and in human consciousness, is towards more and greater ordered coherence. Thus once particles (bosons) are split off from the vacuum's coherent ground state, there is a long, slow process of rediscovering (creatively rediscovering in partnership with fermions) a new coherence.

We human beings, with our need to form a coherent world, do a great deal to further that process of evolving coherence, first as a species, then as individuals, and finally through our relationships and our culture. Each step is an advanced stage in creating greater ordered coherence. It is possible to speculate that at each stage of its evolution, this process would itself be in dialogue with the vacuum (which we

---

*At least all conscious potentiality—the world of matter may arise from an incoherent field within the vacuum.

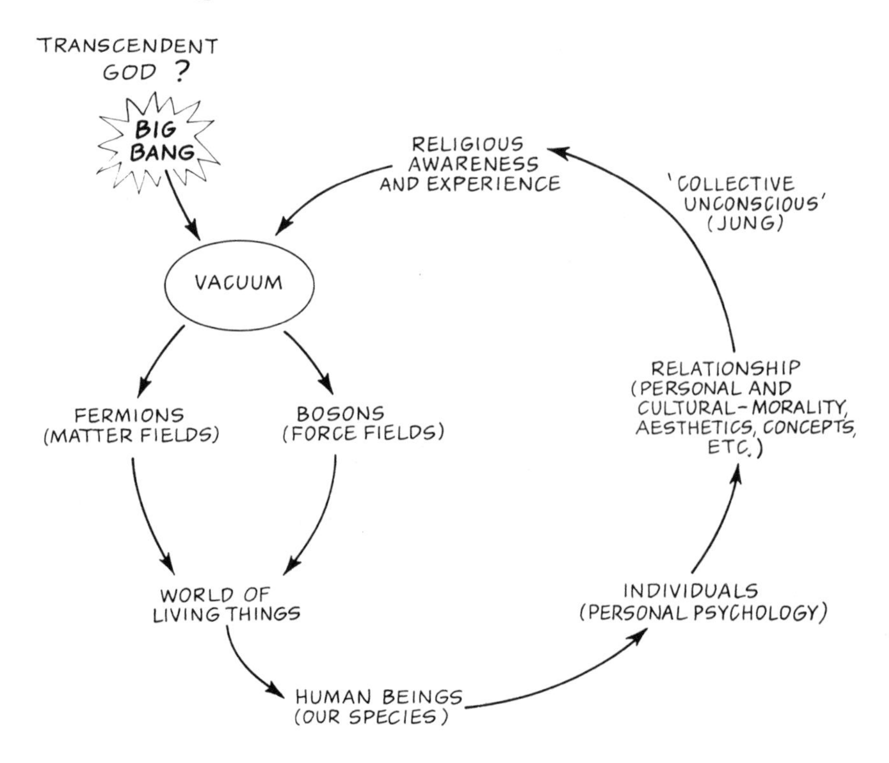

Figure 15.2. The chain of evolving consciousness. Things emerge as fluctuations (excitations) in the vacuum, grow towards renewed coherence, and return to the vacuum as "enriched" fluctuations.

might call God), leading to further fluctuations within it. Mystical experiences are sometimes described as though they might be mirroring such a dialogue (Figure 15.2).

Again, in religious terms, the basic drive towards greater ordered coherence might be seen as the physical basis of Grace, that which allows us, through relationship, to transcend our individuality (the Fall) and to return to unity (God). In Jewish terms, the saving relationship is the People Israel (the Law); for Christians, the Body of Christ. In broader quantum terms, it is the process of overlapping with and coming into nonlocal correlation with each other (and each other's worlds) as fellow quantum systems—seeing, feeling, and becoming part of this process (Figure 15.3).

Jung tells of a belief held by America's Pueblo Indians that they are sons of the Sun, and that in this role it is their daily duty to perform a ritual that helps Father Sun to traverse the sky. They feel this duty as an awesome responsibility and feel that they carry it out for the benefit of the whole world. Jung says of this belief:

> I then realized on what the "dignity," the tranquil composure of the individual Indian, was founded. It springs from his being a son of the

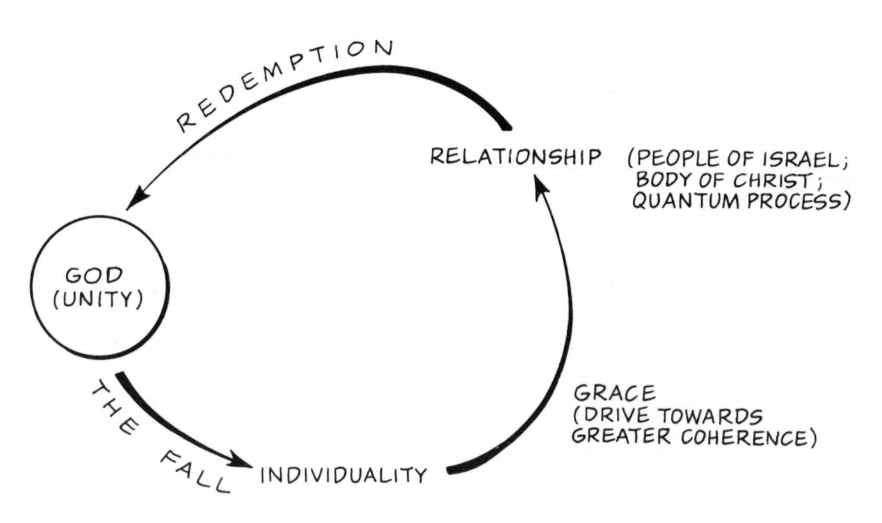

Figure 15.3. The evolution of consciousness seen in religious categories.

> sun; his life is cosmologically meaningful, for he helps the father and preserver of all life in his daily rise and descent. If we set against this our own self-justifications, the meaning of our own lives as it is formulated by our reason, we cannot help but see our poverty.[16]

Understanding the basic origins of consciousness and our own place within its evolution might help us to transcend that poverty.

Because the whole evolutionary process being described is a quantum process, it might be expected to have many "virtual transitions," or probabilistic "trial runs." The process that leads to us as a crucial link in the circle of evolving consciousness (evolving ordered coherence) may not in the end be the one that survives. Our success as lasting partners in the drama of evolution would be subject to the same "constraint of the natural" as the success, or lack of it, of our morality of our aesthetics. If our existence leads to more and greater ordered coherence within the universe, we will succeed as a species; if not, we will fail. We are, in the meantime, a trial run, a probabilistic ripple on the pond, but even as such, we will leave our mark. Remember David Bohm's remark that virtual transitions have many real effects.

Stephen Hawking has said that if we were to discover a complete theory of cosmology, we might come to know the mind of God.[17] I would suggest that if we truly understood our place in the evolving universe, we might come to see ourselves as thoughts (excitations) in the mind of God. In some very important sense, each of us lives his life within a cosmic context.

CHAPTER 16

THE QUANTUM WORLD VIEW

> We are the music makers,
>   And we are the dreamers of dreams. . . .
> Yet we are the movers and shakers
>   Of the world forever it seems.
>
> We, in the ages lying,
>   In the buried past of the earth,
> Built Nineveh with our sighing,
>   And Babel itself with our mirth;
> And o'erthrew them with prophesying
>   To the old of the new world's worth;
> For each age is a dream that is dying,
>   Or one that is coming to birth.
>
> —Arthur William Edgar O'Shaughnessy
> "Ode"

In O'Shaughnessy's poem, the music makers and the dreamers of dreams are thought to be the poets, writers, and philosophers who give us visions, people themselves gifted with some larger-than-life ability to sense and express the dream that is any age.

But in some very meaningful way, each of us, by the very nature of our consciousness and the need of that consciousness to integrate its experience, is a visionary on at least some small scale. Each time that a child makes a clay pot or a person makes a decision, each in some sense creatively discovers some element of the vision that unites us all—our world view. The more so when the child fits together the pieces of his world or the person wonders about the meaning of his life. Each of us is, as Rilke says, a bee of the invisible. That is the meaning of our creativity, and its awesome responsibility.

The child may not be able to articulate fully the way his world fits together, nor the man express the meaning of his life in so many words. For most of us, a world view is a *lived* truth, something we just take for granted and seldom try to describe. Indeed, there is normally motivation to do so only if something goes wrong, if in some way our world view is inadequate or is changing. Only then do we become self-conscious about it.

At the most personal level, a world view is a theme running through a life, a thread that draws apparently disparate pieces together and joins them into a coherent whole. Each of us has, or at least strives to acquire, one. We search for the pattern in terms of which the decisions that we make or the actions we carry out make sense. We ask how our adult lives relate to those of our childhood, how our achievements relate to our youthful aspirations or parental expectations, how our pastimes and acquisitions and our work relate to what we value.

If, at this personal level, one fails to sense some coherent world view, then life itself fragments. We say that such a person has "lost his sense of direction" or "doesn't know who he is." The alienation suffered at this level is alienation from the self.

At a more social level, a world view draws together the many elements of our relationships to others, both the intimate relationships in terms of which we are largely defined and the more general group and social relationships that form some important part of us—our circle of friends or workmates, our neighbors, our "crowd" (those with whom we share an enthusiasm or an interest), our country, and our culture. All of us ask how the individuals that we feel ourselves to be relate to the activities, concerns, and expectations of others. We look at a loved one and see how his aspirations relate to our own; we see a badge or a flag or a painting, hear a chant or an anthem or a piece of music, and sense some response in ourselves. We make these things our own because they express what we are.

If, at this social level, one fails to sense some coherent world view, the sense of self and others breaks down. Both the feeling of belonging and the morality that follows naturally from such a feeling fragment. We feel that we are loners, outsiders, or misfits. The alienation suffered at this level is alienation from society, in its broadest sense.

At a more general level still, a world view is a theme which integrates the sense of self, the sense of self and others, and the sense of how these relate to the wider world—to Nature and other creatures, to the environment as a whole, to the planet, the universe, and ultimately to God—to some overall purpose or sense of direction.

It is at this level that all of us ask why we were born and why we must die, what the meaning of our lives and pursuits are, what *good* we are doing or what sense our suffering has, and where we fit into the general scheme of things.

If, at this most general level, one fails to sense some coherent world view, the sense of self and world disintegrates. We feel that we are "empty," that our lives are "pointless" or "absurd," that "it's all for nothing." The alienation suffered at this level is a general spiritual alienation.

A successful world view must, in the end, draw all these levels—the personal, the social, and the spiritual—into one coherent whole. If it does so, the individual has access to some sense of who he is, why he is here, how he relates to others, and how it is valuable to behave. If it does not, the world it was meant to articulate will fragment and the individual will suffer alienation on some level, perhaps on all levels.

The success or failure of a world view, the birth of a new world view where an old one has failed, rests, ultimately, with the individual and the extent to which he is in touch with his own experience and his own deepest intuitions. As Jung said, "In the last analysis, the essential thing is the life of the individual. . . . We make our own epoch."[1] An important part of the experience out of which the individual will make his epoch will be his knowledge about the world and himself.

For the better part of the past two thousand years, the vast majority of people in the West successfully embraced the Judeo-Christian world view, whether they were members of a religious group or not. Most obviously this world view offered some sense of how the individual related to the cosmos and to Nature. Human beings were the transcendent God's special creation, made in the image of that God. We lived our lives within a God-given order and had been given dominion over all the earth. At the end of our days we would experience some sort of judgment, rebirth, or afterlife.

But the God-given order related not just to the cosmos but to nearly every detail of social and personal life. Through belief in God and acceptance of His Law or Son, the individual knew how to behave and feel towards others, how to conduct his business or educate his children, how to build his buildings or write his music, even how and when and within what framework to make love. There were many trivial, daily exceptions, but each individual had a unifying theme running through his life at every level, a sense of who he was and where he belonged and why his life had meaning.

The traditional Judeo-Christian world view started to lose its coherence only when the discoveries of modern science began to undermine many of the cosmological assumptions on which it rested. The individual's growing knowledge about himself and his world no longer accorded with the main features of the biblical creation story, with the cosmology of an earth-centered universe and the biology of human uniqueness, nor with the spirit of physics-defying miracles, heavenly messengers, and divine intervention.

The new spirit of the age was to *understand,* to say rationally how one thing followed from another and to explain the exact mechanics by which it did so. The mechanical world view was born and, as I have already argued at length in this book, touched nearly every aspect of modern life.

But where the Judeo-Christian world view was a success because it drew every aspect of the individual and every level of his life into one coherent whole, the mechanical world view could never really succeed. From the beginning, it was flawed by its inability to explain or account for consciousness. It suffered in consequence from what American philosopher Lawrence Cahoone calls "the three pernicious dichotomies"—the split between subject and object (mind and body, inner and outer), the split between the individual and his relationships, and the split between the world of human culture and the natural realm of biophysical processes.[2]

The mechanical world view successfully gave us a science that explained *things,* and a technology to exploit them as never before, but the price paid was a kind of alienation at every level of human life.

The three "pernicious dichotomies" left us wondering how we conscious human beings related to ourselves (our own bodies, our own pasts and futures, our own subselves), to each other, and to the world of Nature and facts. In trying to resolve these questions, our psychology, our philosophy, and our religion fragmented into opposite extremes. As Yeats said of this era, "Things fall apart; the center cannot hold."[3]

The split between mind and body, or between inner and outer, gave rise to the dichotomy between extreme subjectivism (a world without objects) and extreme objectivism (a world without subjects). Thus idealism denied the reality or importance of matter and reduced everything to mind, while materialism denied the reality or importance of mind and reduced everything to matter. Freud assumed that the inner was real and accessible, while the outer was all projection, and many strains of mysticism mirrored this view—e.g., proclaiming that the world is the veil of Maya, a veil of illusion. At the other extreme, behaviorism assumed the outer was real but denied the relevance of the inner. It became psychology without the psyche.

The split between the individual and his relationships led on the one hand to an exaggerated individualism, to a selfish will to power and possession, and on the other to an enforced communitarianism like that of Marxism, which denies that individuals have any meaning or importance at all while stressing the absolute primacy of relationship.

The split between culture and Nature led both to relativism of all sorts—factual, moral, aesthetic, and spiritual (value judgments)—and to dogma and extreme fundamentalism. There seemed no middle ground between the two extremes of saying that a given way of looking at things was only one of many contingent and relative ways of looking at them, or saying there was only one true and absolute way of looking at them. No way to say that we were not either wholly creatures of culture, and therefore unrooted in any established facts, or wholly creatures of Nature (of the given), with no flexibility or room for creative development.

In the West, these dichotomies robbed our individuality of its context and landed us in the deepest isolation, leading to narcissism. They left us cut off from an outer confirmation of our inner life, leading to nihilism; and denied us the confirmation of our ideas, leaving us with relativism and subjectivism. Each nourished a form of alienation, and the sum total of this alienation is the curse of modernism.

The mechanical world view fails, ultimately, because it does not work towards a greater, ordered coherence. It reflects neither the intuitions nor the personal needs of most people, nor the simple, quite classical fact that we live in a shrinking world, a world where technology and mass communication, industrial pollution and the threat of global extinction have made us aware as never before that in some very important sense we are all interdependent and that our human lives are inseparably intertwined with the world of Nature. A world view that leads to fragmentation, and that encourages the selfish exploitation of others and of our common world, violates the constraint of the natural. It lessens rather than increases coherence.

The mechanical world view, as I have argued, owes most to the dualist philosophy of Descartes and the mechanistic physics of Newton. In recent years, many people have begun to sense that the new physics, primarily quantum physics, holds out the promise of a new world view, one that would give some physical basis to a more holistic, less fragmented way of looking at ourselves within the world.

Hence the many books and articles written about quantum physics and holism, quantum physics and Eastern mysticism, quantum physics and healing, quantum physics and psychic phenomena, and so on. All have been partial and groping attempts to articulate something that is in the air, something that answers people's need for a more coherent world picture—a need to find a unifying explanation of ourselves and our universe and a unifying foundation for our behavior. But none has really grounded this need itself in the actual physics of consciousness, and thus none has been able to lay a solid, physical basis for a quantum world view.

Once we have made this connection, once we have seen that the physics of human consciousness emerges from quantum processes within the brain and that in consequence human consciousness and the whole world of its creation shares a physics with everything else in this universe—with the human body, with all other living things and creatures, with the basic physics of matter and relationship, and with the coherent ground state of the quantum vacuum itself—it becomes impossible to imagine a single aspect of our lives that is not drawn into one coherent whole.

The quantum world view transcends the dichotomy between mind and body, or between inner and outer, by showing us that the basic building blocks of mind (bosons) and the basic building blocks of matter (fermions) arise out of a common quantum substrate (the vacuum) and are engaged in a mutually creative dialogue whose roots can be traced back to the very heart of reality creation. Crudely put, mind is relationship and matter is that which it relates. Neither, on its own, could evolve or express anything; together they give us ourselves and the world.

The creative dialogue between mind and matter is the physical basis of all creativity in the universe and is also the physical basis of human creativity. The quantum self experiences no dichotomy between the inner and the outer because the two, the inner world of mind (of ideas, values, notions of goodness, truth, and beauty) and the outer world of matter (of facts), give rise to each other.

The quantum world view transcends the dichotomy between the individual and relationship by showing us that people can only be the individuals they are within a context. I am my relationships—my relationships to the subselves within my own self (my past and my future), my relationships to others, and my relationships to the world at large.

I am I, uniquely myself, because I am an utterly unique pattern of relationships, and yet I cannot separate this I who I am from those relationships. For the quantum self, neither individuality nor relationship is primary because both arise simultaneously and with equal "weight" from the quantum substrate. In the case of individual persons and their relationships, that substrate is a Bose-Einstein condensate in the brain; in the case of individual particles and their relationships, that substrate is a Bose-Einstein condensate in the quantum vacuum.

The quantum self thus mediates between the extreme isolation of Western individualism and the extreme collectivism of Marxism or Eastern mysticism.

Similarly, the quantum world view transcends the dichotomy between human culture and Nature, and indeed imposes the constraint of the natural upon the ultimate success of the cultural.

The physics of consciousness that gives rise to the world of culture—to art, ideas, values, moralities, and even to religions—is the same physics that gives us the world of Nature. In both cases it is a physics driven by the need to maintain and increase ordered coherence in free response to the environment. The quantum self is, by the very mechanics of its consciousness, a natural self—a free and responsive self—and its world, ultimately, will reflect the world of Nature. Where it does not, that world will fail.

In summary, the quantum world view stresses dynamic relationship as the basis of all that is. It tells us that our world comes about through a mutually creative dialogue between mind and body (inner and outer, subject and object), between the individual and his personal and material context, and between human culture and the natural world. It gives us a view of the human self that is free and responsible, responsive to others and to its environment, essentially related and naturally committed, and at every moment creative.

Notes

Preface

1. Danah Zohar, 1982.
2. Christopher Rawlence, 1985.

Chapter 1 A Physics of Everyday Life

1. Bertrand Russell, 1957, p. 45.
2. Michel Serres, quoted in Ilya Prigogine and Isabelle Stengers, 1984, pp. 304–305.
3. Jacques Monod, 1972.
4. Michel Serres, quoted in Ilya Prigogine and Isabelle Stengers, 1984, pp. 304–305.

Chapter 2 What's New About the New Physics?

1. Quoted in Arthur Fine, 1986, dust jacket.
2. Quoted in Arthur Koestler, 1972, p. 52.
3. Dr. Ian Aitchison, Department of Theoretical Physics, University of Oxford, lecture comment.
4. Bernard d'Espagnat, 1979, p. 128.
5. David Bohm, 1951, p. 415.
6. Ibid., p. 414.
7. R. L. Pfleegor and L. Mandel, 1967.

Chapter 3 Consciousness and the Cat

1. J. S. Bell, 1987, Chapter 20.
2. N. Mermin, 1985, p. 38.
3. Ilya Prigogine, 1980, pp. 241–248; Roger Penrose, 1986, Chapter 4.
4. Abner Shimony, 1988, pp. 36–43.
5. John Archibald Wheeler, 1980. David Bohm (1986, p. 126) also uses the expression *participative universe,* and Ilya Prigogine (Prigogine and Isabelle Stengers, 1984, p. 299) speaks of "knowledge that is both objective and participatory."

6. John Archibald Wheeler, 1983, p. 199.
7. Quoted ibid., p. 197.
8. Ilya Prigogine and Isabelle Stengers, 1984, p. 293.
9. M. Merleau-Ponty, 1960, pp. 136–137.
10. This problem is discussed at length in Allan Bloom, 1987, p. 160 et passim.
11. Fritjof Capra, 1983, p. 77.
12. Werner Heisenberg, 1958.

Chapter 4 Are Electrons Conscious?

1. Genesis 1:26.
2. John 18:36.
3. J. B. Watson, 1913, pp. 163, 176.
4. Thomas Nagel, 1979, pp. 165–180.
5. W. H. Thorpe, 1974, pp. 44–45.
6. Ibid., p. 45.
7. N. P. Franks and W. R. Lieb, 1988, p. 662.
8. Heraclitus, Fragments 38 and 45, Joh Burnet, 1963, p. 136.
9. Arthur O. Lovejoy, 1964.
10. Paul Edwards, 1967, "Panpsychism."
11. Ibid.
12. Ibid.
13. J. E. Lovelock, 1982.
14. Thomas Nagel, 1979, p. 181.
15. Ibid., p. 187.
16. Karl R. Popper and John C. Eccles, 1977, p. 11.
17. Thomas Nagel, 1979, p. 184.
18. David Bohm, 1986, p. 129.
19. Ibid., p. 122.

Chapter 5 Consciousness and the Brain

1. Colin McGinn, 1987, p. 283.
2. Joseph Weizenbaum, 1984, p. 6.
3. Ibid.
4. A. N. Whitehead, 1979, p. 109.
5. Ann Treisman, 1986, pp. 106–115.
6. Ibid., p. 115.
7. René Descartes, 1960, p. 86.
8. For instance, see Thomas Nagel, 1986, p. 50, or Hubert L. Dreyfus and Stuart E. Dreyfus, 1988, pp. 58–63.

9. Ken Wilber, 1982, p. 2.
10. Ibid., p. 7.
11. David Bohm, 1980.
12. Quoted in Ken Wilber, 1982, p. 25.
13. Ken Wilber, 1982, p. 2.
14. Daniel Dennett, 1984, p. 1453.

Chapter 6 A Quantum Mechanical Model of Consciousness

1. Much of the material in this chapter is based upon I. N. Marshall, 1989.
2. David Bohm, 1951, p. 169.
3. Ibid., p. 170.
4. Roger Penrose, 1987, p. 274.
5. I. N. Marshall, 1960.
6. Yuri Orlov, 1982, p. 45.
7. John Crook, 1987, p. 389.
8. For example, Evan Harris Walker, 1970.
9. H. Fröhlich, 1968.
10. H. Fröhlich, 1986.
11. H. Fröhlich and F. Kremer, 1983, p. 1.
12. Lawrence Domash, 1976, p. 657.
13. Fritz-Albert Popp, 1988, pp. 576–585.
14. Humio Inaba, Tohoku University Research Institute of Electrical Communication in Sendai, Japan, 1989, p. 41.
15. W. B. Chwirot, 1986, pp. 821–886.
16. M. Rattemeyer and F.-A. Popp, 1981.
17. I. N. Marshall, "Excitations of a Bose-Einstein Condensate," forthcoming.
18. Ibid.
19. R. W. Thatcher and E. R. John, 1977.
20. Michael P. Stryker, 1989, pp. 297–298.
21. W. R. Adey, 1980, pp. 119–125; I. N. Marshall, "Excitations of a Bose-Einstein Condensate," forthcoming.

Chapter 7 Mind and Body

1. Romans 7:24–25.
2. Plato, *Phaedo,* 66a–e.
3. M. Jammer, 1967.
4. Thomas Nagel, 1986, p. 28.
5. Quoted in Gordon G. Globus et al., 1977, p. 320.
6. Colin McGinn, remark made in Mind/Body Seminar, Oxford, 1988.

7. Paul Teller, 1986, pp. 71–81.
8. Plato, *Timaeus,* 31c.
9. Martin Buber, 1937, p. 14.
10. Martin Heidegger, 1964, p. 680.
11. Ibid., p. 684.
12. Lucretius, 1986, Book III.

Chapter 8 The Person That I Am

1. Cited in Derek Parfit, 1984, p. 245.
2. Thomas Nagel, 1979, p. 164.
3. Ibid., p. 164.
4. Derek Parfit, 1984, pp. 273, 275.
5. An expression used by Derek Parfit. See 1984, pp. 245–280.
6. Martin Heidegger, 1962, p. 329.
7. Derek Parfit, 1984, p. 281.
8. Gary Zukav, 1979, p. 96.
9. Arthur Lovejoy, 1964, p. 29.
10. Erich Neumann, 1954, Part I.
11. Derek Parfit, 1975, pp. 218–219.
12. T. S. Eliot, "Burnt Norton," *Four Quartets.*

Chapter 9 The Relationships That I Am

1. Sigmund Freud, 1915.
2. Rosemary Gordon, 1965.
3. Martin Buber, 1937, p. 5.
4. Ibid., p. 4.
5. Martin Heidegger, 1962, pp. 160, 163.
6. Jean-Paul Sartre, 1956, p. 282.
7. Charles Rycroft, 1968, p. 101.
8. Melanie Klein, "The Emotional Life of the Infant," quoted in Rosemary Gordon, 1965, p. 128.
9. R. L. Pfleegor and L. Mandel, 1967.
10. Erik Erikson, 1963.
11. The realization that neural pathways are laid down as we grow rather than all being present at birth is a fairly recent breakthrough in neural research. See, for instance, Colin Blakemore, 1988.
12. Robert Lindner, 1962, pp. 196–201.
13. H. D. Zeh, 1983, p. 346.
14. Ibid., p. 346.
15. See, for instance, S. H. Foulkes and E. J. Anthony, 1957, or E. Bruce Taub-Bynum, 1984.

Chapter 10 The Survival of the Self

1. Anthony Flew, 1967, p. 141.
2. Gabriel Marcel, 1951, p. 148.
3. Gabriel Marcel, 1964, p. 150.
4. Gabriel Marcel, 1960, p. 242.
5. T. S. Eliot, "East Coker," *Four Quartets.*
6. R. L. Pfleegor and L. Mandel, 1967.

Chapter 11 Getting Beyond Narcissism

1. Frederick Perls, 1969, epigraph.
2. Christopher Lasch, 1979, p. 72.
3. Dr. Malcolm Pines, London.
4. Richie Herink, 1980.
5. Jerome D. Frank, 1975.
6. V. Mansfield and J. M. Spiegelman, 1989.
7. Allan Bloom, 1987, p. 125.
8. Ibid., p. 164.
9. V. Mansfield and J. M. Spiegelman, 1989, pp. 3–31.
10. Charles Rycroft, 1968.
11. Ibid., p. 101.
12. Sigmund Freud, 1962, p. 62.
13. Jean-Paul Sartre, 1956, p. 569.
14. Jean-Paul Sartre, 1957, pp. 22–23.
15. Christopher Lasch, 1979, pp. 23, 25.
16. Martin Heidegger, 1964, pp. 680, 684.
17. R. M. Wald, 1986.
18. Plato, *Republic,* Book III.
19. John Donne, *Devotions,* XVII.
20. Arthur Miller, 1989.
21. Jean-Paul Sartre, 1957, p. 22.
22. Sigmund Freud, 1962, p. 90.
23. C. G. Jung, 1964, p. 154.

Chapter 12 The Free Self

1. Aristotle, *Poetics,* 10.
2. Martin Luther, 1961, iv.
3. Bertrand Russell, 1957, p. 54.
4. Charles Rycroft, 1966, p. 13.
5. Ibid., p. 12.
6. Richard Taylor, 1967.

7. James Gleick, 1988.
8. For instance, John Lucas, 1970, pp. 107–113, or Fred Alan Wolf, 1981, pp. 234–241.
9. T. D. Clark, 1987.
10. David Deutsch, 1985, pp. 97–117.
11. S. W. Kuffler and J. G. Nicholls, 1976.
12. Charles Rycroft, 1966, pp. 7–22.
13. Jean-Paul Sartre, 1962, pp. 122–125.
14. Ibid., p. 123.
15. Charles Taylor, 1985, p. 35.

Chapter 13 The Creative Self

1. Gershom Scholem, 1983, "Creation of Man."
2. T. A. Goudge, 1967, p. 293.
3. Rainer Maria Rilke, "Ninth Elegy," in Gabriel Marcel, 1951, p. 257.
4. Ilya Prigogine and Isabelle Stengers, 1984.
5. William Wordsworth, *The Prelude,* Book II, lines 273–275 (1805–6 version).
6. John Cairns, et al., 1988, pp. 142–145.
7. K. H. Li, F.-A. Popp, et al., 1983, pp. 117–122.
8. Ilya Prigogine and Isabelle Stengers, 1984, p. 301.
9. Fritz-Albert Popp, remark made at Oxford Conference on Biological Coherence, May 1989.
10. W. H. Thorpe, 1963, pp. 201–202.
11. I. N. Marshall, "Excitations of a Bose-Einstein Condensate," forthcoming.
12. David Bohm, 1951, p. 415.
13. Charles Taylor, 1985, p. 36.
14. C. G. Jung, 1964, Vol. 10, p. 149.
15. Jean-Paul Sartre, quoted in Roger Scruton, 1981, p. 268.
16. Lawrence E. Cahoone, 1988, pp. 215–216.
17. Ilya Prigogine and Isabelle Stengers, 1984, p. 299.
18. Ibid., p. 298.
19. Charles Taylor, 1985, p. 44.

Chapter 14 Ourselves and the Material World

1. In the discussion about the body and its needs that follows, it would be impossible in many places for me to distinguish between my own ideas and those of Samuel Todes, to say exactly where or when I am unwittingly quoting from him. His own ideas are published in Todes, 1989.

2. Jacques Ellul, 1967, pp. 30–31.
3. Henry Adams, 1961.
4. Ilya Prigogine and Isabelle Stengers, 1984.
5. Martin Heidegger, 1964, pp. 649–701.
6. Ibid.
7. Rainer Maria Rilke, *Selected Letters,* quoted in Gabriel Marcel, 1951, p. 258.

Chapter 15 The Quantum Vacuum and the God Within

1. Brian Pippard, 1986, p. 555.
2. Rodney Loudon, 1983, Chapter 6.
3. Fritz-Albert Popp, remark made at Oxford Conference on Biological Coherence, May 1989.
4. Gregoire Nicolis, 1989, p. 330.
5. I. N. Marshall, "Identity, Coalescence and Collapse in Quantum Mechanics," forthcoming.
6. John D. Barrow and Frank J. Tipler, 1988.
7. Ibid., p. 22.
8. Ilya Prigogine and Isabelle Stengers, 1984, p. 301.
9. Tony Hey and Patrick Walters, 1989, p. 130.
10. David Finkelstein, 1989, p. 1.
11. "Imagine that the 'vacuum' in which we live is analogous to a 'weak superconductor.' . . ." Tony Hey and Patrick Walters, 1989, p. 151. Also, David Finkelstein, 1989, p. 7.
12. I. N. Marshall, "Excitations of a Bose-Einstein Condensate," forthcoming.
13. Pierre Teilhard de Chardin, 1959.
14. John B. Cobb, Jr., and David Ray Griffin, 1976.
15. Pierre Teilhard de Chardin, 1959, p. 221.
16. C. G. Jung, 1963, pp. 236–237.
17. Stephen Hawking, 1988, p. 175.

Chapter 16 The Quantum World View

1. C. G. Jung, p. 149.
2. Lawrence E. Cahoone, 1988, pp. 233–234.
3. "The Second Coming."

Bibliography

Adams, Henry. *Mount-Saint-Michel and Chartres.* New York: New American Library, 1961.

Adey, W. R. "Frequency and Power Windowing in Tissue Interactions with Weak Electromagnetic Fields." *Proceedings of the IEEE,* Vol. 68, No. 1 (1980).

Aristotle. *Basic Works,* ed. Richard McKeon. New York: Random House, 1941.

Augustine, Saint. *Confessions.* Harmondsworth, UK: Penguin Books, 1961.

Barrow, John D., and Tipler, Frank J. *The Anthropic Cosmological Principle.* Oxford and New York: Oxford University Press, 1988.

Bell, J. S. *Speakable and Unspeakable in Quantum Mechanics.* Cambridge: Cambridge University Press, 1987.

Blakemore, Colin. *The Mind Machine.* Oxford: Basil Blackwell, 1988.

Blakemore, Colin, and Greenfield, Susan, eds. *Mindwaves.* Oxford: Basil Blackwell, 1987.

Bloom, Allan. *The Closing of the American Mind.* New York and London: Simon & Schuster, 1987.

Bohm, David. *Quantum Theory.* London: Constable, 1951.

Bohm, David. *Wholeness and the Implicate Order.* London, Boston, and Henley, UK: Routledge & Kegan Paul, 1980.

Bohm, David. "A New Theory of the Relationship of Mind and Matter." *The Journal of the American Society for Psychical Research,* Vol. 80, No. 2 (1986).

Bohr, Niels. "On Atoms & Human Knowledge." *Daedalus,* Vol. 87, No. 2 (1958).

Buber, Martin. *I and Thou.* Edinburgh: T. & T. Clark, 1937.

Burnet, Joh. *Early Greek Philosophy.* Cleveland and New York: Meridian Books, 1963.

Cahoone, Lawrence E. *The Dilemma of Modernity.* Albany: State University of New York Press, 1988.

Cairns, John, et al. "The Origin of Mutants." *Nature,* Vol. 335 (8 September 1988).

Capra, Fritjof. *The Turning Point.* London: Fontana Flamingo, 1983.

Chwirot, W. B. "New Indication of Possible Role of DNA in Ultraweak Photon Emission from Biological Systems." *Journal of Plant Physiology,* Vol. 122 (1986).

Clark, T. D. "Macroscopic Quantum Objects." In B. J. Hiley and F. David Peat, eds., *Quantum Implications.* London and New York: Routledge & Kegan Paul, 1987.

Cobb, John B., Jr., and Griffin, David Ray. *Process Theology: An Introductory Exposition.* Philadelphia: Westminster Press, 1976.

Crick, F.H.C. "Thinking About the Brain." *Scientific American,* Vol. 241, No. 3 (1979).

Crook, John. "The Nature of Conscious Awareness," In Colin Blakemore and Susan Greenfield, eds., *Mindwaves.* Oxford: Basil Blackwell, 1987.

Dennett, Daniel C. "Computer Models and the Mind—A View from the East Pole." *Times Literary Supplement,* 14 December 1984.

Descartes, René. *Meditations.* Indianapolis and New York: Bobbs-Merrill Co., 1960.

Descartes, René. *Essential Works of Descartes.* New York: Bantam, 1961.

d'Espagnat, Bernard. "The Question of Quantum Reality." *Scientific American,* Vol. 241 (November 1979).

Deutsch, David. "Quantum Theory, the Clark-Turing Principle and the Universal Quantum Computer." *Proceedings of the Royal Society of London,* A400 (1985).

Domash, Lawrence H. "The Transcendental Meditation Technique and Quantum Physics: Is Pure Consciousness A Macroscopic Quantum State In the Brain?" In David Orme-Johnson and John T. Farrow, eds., *Scientific Research on the Transcendental Meditation Program,* Vol. I. Geneva: Maharishi European Research University Press, 1976.

Donne, John. *Devotions.* 1624.

Dreyfus, Hubert L., and Dreyfus, Stuart E. *Mind over Machine.* New York: Free Press, Macmillan, 1988.

Edwards, Paul, ed. *The Encyclopedia of Philosophy.* London: Collier-Macmillan, 1967.

Edwards, Paul. "Panpsychism." In Paul Edwards, ed., *The Encyclopedia of Philosophy.* London: Collier-Macmillan, 1967.

Ellul, Jacques. *The Technological Society.* New York: Vintage Books, 1967.

Erikson, Erik H. *Childhood and Society.* New York: W. W. Norton & Co., 1963.

Feigl, Herbert. *The Mental and the Physical.* Minneapolis: University of Minnesota Press, 1959.

Fine, Arthur. *The Shaky Game.* Chicago and London: University of Chicago Press, 1986.

Finklestein, David. "A Theory of the Vacuum." In S. Saunders, ed., *Philosophy of the Vacuum.* Oxford: Oxford University Press, 1989.

Flew, Anthony. "Immortality." In Paul Edwards, ed., *The Encyclopedia of Philosophy.* London: Collier-Macmillan, 1967.

Foulkes, S. H., and Anthony, E. J. *Group Psychotherapy.* Harmondsworth, UK: Penguin Books, 1957.

Frank, Jerome D. "An Overview of Psychotherapy." In Gene Usdin, ed., *Overview of the Psychotherapies,* New York: Bruner/Mazel, 1975.

Franks, N. P., and Lieb, W. R. "Volatile General Anaesthetics Activate a Novel Neuronal K^+ Current." *Nature,* Vol. 333 (16 June 1988).

Freedman, S., and Clauser, J. "Experimental Test of Local Hidden Variables Theories." *Physical Review Letters,* Vol. 28 (1972).

Freud, Sigmund. *Instincts and Their Vicissitudes* (1915), Standard Edition, Vol. 14. London: Hogarth Press, 1957.

Freud, Sigmund. *Civilization and its Discontents.* New York: W. W. Norton & Co., 1962.

Fröhlich, H. "Long-Range Coherence and Energy Storage in Biological Systems." *International Journal of Quantum Chemistry,* Vol. II (1968).

Fröhlich, H. "Coherent Excitations in Active Biological Systems." In F. Gutman and H. Keyzer, eds., *Modern Bioelectrochemistry.* New York and London: Plenum, 1986.

Fröhlich, H., and Kremer, F., eds. *Coherent Excitations in Biological Systems.* Berlin, Heidelberg, New York, and Tokyo: Springer-Verlag, 1983.

Gleick, James. *Chaos.* London: Heinemann, 1988.

Globus, Gordon G., et al., eds. *Consciousness and the Brain.* New York and London: Plenum, 1977.

Gordon, Rosemary. "The Concept of Projective Identification." *Journal of Analytical Psychology,* Vol. 10, No. 2 (1965).

Goudge, T. A. "Henri Bergson." In Paul Edwards, ed., *The Encyclopedia of Philosophy.* London: Collier-Macmillan, 1967.

Hawking, Stephen. *A Brief History of Time.* London and New York: Bantam, 1988.

Heidegger, Martin. *Being and Time.* New York and Evanston, Ill.: Harper & Row, 1962.

Heidegger, Martin. "The Origin of the Work of Art." In Albert Hofstadter and Richard Kuhns, eds., *Philosophies of Art and Beauty.* New York: Modern Library, 1964.

Heisenberg, Werner. "The Representation of Nature in Contemporary Physics." *Daedalus,* Vol. 87 (1958).

Herbert, Nick. *Quantum Reality.* London: Rider & Co., 1985.

Herink, Richie, ed. *The Psychology Handbook: The A to Z Guide to More than 250 Different Therapies in Use Today.* New York and London: New American Library, 1980.

Hey, Tony, and Walters, Patrick. *The Quantum Universe.* Cambridge: Cambridge University Press, 1989.

Hill, Carol. *The Eleven Million Mile High Dancer.* New York and London: Penguin Books, 1986.

Inaba, Humio. Research reported in *The New Scientist,* 27 May 1989.

Jammer, M. "Mass." In Paul Edwards, ed., *The Encyclopedia of Philosophy.* London: Collier-Macmillan, 1967.

Johnson, Stephen M. *Humanizing the Narcissistic Style.* New York and London: W. W. Norton & Co., 1987.

Jung, C. G. *Memories, Dreams, Reflections.* London: Collins, Routledge & Kegan Paul, 1963.

Jung, C. G. "The Meaning of Psychology for Modern Man." In Sir Herbert Read et al., eds., *C. G. Jung: The Collected Works,* Vol. 10. London: Routledge & Kegan Paul, 1964.

Koestler, Arthur. *The Roots of Coincidence.* London: Hutchinson, 1972.

Kuffler, S. W., and Nicholls, J. G. *From Neuron to Brain.* Sunderland, Mass.: Sinauer, 1976.

Lasch, Christopher. *The Culture of Narcissism.* New York: Warner Books, 1979.

Li, K. H.; Popp, F.-A., et al. "Indications of Optical Coherence in Biological Systems and Its Possible Significance." In H. Fröhlich and F. Kremer, eds., *Coherent Excitations in Biological Systems.* Berlin, Heidelberg, New York, Tokyo: Springer-Verlag, 1983.

Lindner, Robert. *The Fifty Minute Hour.* London: Transworld Publishers, 1962.

Loudon, Rodney. *The Quantum Theory of Light.* Oxford: Clarendon Press, 1983.

Lovejoy, Arthur O. *The Great Chain of Being.* Cambridge, Mass., and London: Harvard University Press, 1964.

Lovelock, J. E. *Gaia.* Oxford: Oxford University Press, 1982.

Lucas, J. R. *The Freedom of the Will.* Oxford: Clarendon Press, 1970.

Lucretius. *On the Nature of the Universe.* New York and London: Penguin Books, 1986.

Luther, Martin. "The Bondage of the Will." In John Dillenberger, ed., *Martin Luther.* New York: Anchor Doubleday, 1961.

McGinn, Colin. "Could a Machine Be Conscious?" In Colin Blakemore and Susan Greenfield, eds., *Mindwaves.* Oxford: Basil Blackwell, 1987.

Mansfield, V., and Spiegelman, J. M. "Quantum Mechanics and Jungian Psychology: Building a Bridge." *Journal of Analytical Psychology,* Vol. 34 (1989).

Marcel, Gabriel. *Homo Viator.* London: Victor Gollancz, 1951.

Marcel, Gabriel. *The Mystery of Being.* Chicago: Henry Regnery Company, 1960.

Marcel, Gabriel. *Creative Fidelity.* New York: Noonday Press, 1964.

Marshall, I. N. (Ninian). "ESP and Memory: A Physical Theory." *British Journal for the Philosophy of Science,* Vol. X, No. 40 (1960).

Marshall, I. N. "Consciousness and Bose-Einstein Condensates." *New Ideas in Psychology,* Vol. 7, No. 1 (1989).

Marshall, I. N. "Excitations of a Bose-Einstein Condensate." Forthcoming.

Marshall, I. N. "Identity, Coalescence and Collapse in Quantum Mechanics." Forthcoming.

Merleau-Ponty, M. "Le Philosophe et la Sociologie." *Éloge de la Philosophie.* Paris: Collection Idées, Gallimard, 1960.

Merleau-Ponty, M. *Phenomenology of Perception.* New York: Humanities Press, 1962.

Mermin, N. "Is the Moon There When Nobody Looks?" *Physics Today,* April 1985.

Miller, Arthur. "Miller, Marx and Marilyn." *The Independent* (London), 3 January 1989.

Monod, Jacques. *Chance and Necessity.* London: Collins, 1972.

Nagel, Thomas. *Mortal Questions.* Cambridge: Cambridge University Press, 1979.

Nagel, Thomas. *The View from Nowhere.* Oxford: Oxford University Press, 1986.

Neumann, Erich. *The Origins and History of Consciousness.* New York: Pantheon Books, 1954.

Nicolis, Gregoire. "Physics of Far-from-Equilibrium Systems and Self-organization." In Paul Davies, ed., *The New Physics.* Cambridge: Cambridge University Press, 1989.

Orlov, Yuri F. "The Wave Logic of Consciousness: A Hypothesis." *International Journal of Theoretical Physics,* Vol. 21, No. 1 (1982).

Parfit, Derek. "Personal Identity." In John Perry, ed., *Personal Identity.* Berkeley, London: University of California Press, 1975.

Parfit, Derek. *Reasons and Persons.* Oxford: Oxford University Press, 1984.

Paul, Saint. Romans, New Testament.

Penrose, Roger. "Big Bangs, Black Holes and 'Time's Arrow.'" In R. Flood and M. Lockwood, eds., *The Nature of Time.* Oxford: Basil Blackwell, 1986.

Penrose, Roger. "Minds, Machines and Mathematics." In Colin Blakemore and Susan Greenfield, eds., *MINDWAVES.* Oxford: Basil Blackwell, 1987.

Perls, Frederick S. *Gestalt Therapy Verbatim.* New York: Bantam, 1969.

Pfleegor, R. L., and Mandel, L. "Interference of Independent Photon Beams." *Physical Review,* Vol. 159, No. 5 (25 July 1967).

Pippard, Brian. "God and the Physical Scientist." *Times Literary Supplement,* 23 May 1986.

Plato. *The Collected Dialogues,* ed. Edith Hamilton and Huntington Cairns. Bollingen Series LXXI. New York: Pantheon Books, 1961.

Popp, Fritz-Albert. "On the Coherence of Ultraweak Photonemission from Living Tissues." In C. W. Kilmister, ed., *Disequilibrium and Self-Organization.* Dordrecht and Boston: D. Reidel Publishing Co., 1986.

Popp, Fritz-Albert, et al. "Physical Aspects of Biophotons." In *Experientia,* Vol. 44. Basel: Birkhauser Verlag, 1988.

Popper, Karl R., and Eccles, John C. *The Self and Its Brain.* Berlin, London, and New York: Springer-Verlag, 1977.

Prigogine, Ilya. *From Being to Becoming.* New York: W. H. Freeman and Co., 1980.

Prigogine, Ilya, and Stengers, Isabelle. *Order out of Chaos.* New York and London: Bantam, 1984.

Rattemeyer, M.; Popp, F.-A., et al. "Evidence of Photon Emission from DNA in Living Systems." *Naturwissenschaften,* Vol. 68, No. 5 (1981).

Rawlence, Christopher, ed. *About Time.* London: Cape, 1985.

Russell, Bertrand. "A Free Man's Worship." In *Mysticism and Logic.* New York: Doubleday Anchor, 1957.

Rycroft, Charles. *Psychoanalysis Observed.* London: Constable, 1966.

Rycroft, Charles. *A Critical Dictionary of Psychoanalysis.* London: Thomas Nelson and Sons, 1968.

Sartre, Jean-Paul. *Being and Nothingness.* New York: Philosophical Library, 1956.

Sartre, Jean-Paul. "Existentialism." In *Existentialism and Human Emotions.* New York: Philosophical Library, 1957.

Sartre, Jean-Paul. *The Flies.* In *No Exit and Three Other Plays.* New York: Vintage Books, 1962.

Scholem, Gershom G., ed. *Zohar: The Book of Splendor.* New York: Schocken Books, 1963.

Scruton, Roger. *From Descartes to Wittgenstein: A Short History of Modern Philosophy.* London and Boston: Routledge & Kegan Paul, 1981.

Shimony, Abner. "Meeting of Physics and Metaphysics." *Nature,* Vol. 291 (4 June 1981).

Shimony, Abner. "The Reality of the Quantum World." *Scientific American,* Vol. 258, No. 1 (January 1988).

Sperry, R. W. "Mental Phenomena as Causal Determinants." In Gordon G. Globus, et al., eds., *Consciousness and the Brain.* New York and London: Plenum Press, 1977.

Stryker, Michael P. "Is Grandmother an Oscillation?" *Nature,* Vol. 338 (23 March 1989).

Stuart, C.I.J.M., et al. "Mixed Brain Dynamics: Neural Memory as a Macroscopic Ordered State." *Foundations of Physics,* Vol. 9, Nos. 3 and 4 (1979).

Taub-Bynum, E. Bruce. *The Family Unconscious.* Wheaton, Ill.: Theosophical Publishing House, 1984.

Taylor, Charles. *Human Agency and Language.* Cambridge and New York: Cambridge University Press, 1985.

Taylor, Richard. "Determinism." In Paul Edwards, ed., *The Encyclopedia of Philosophy.* London: Collier-Macmillan, 1967.

Teilhard de Chardin, Pierre. *The Phenomenon of Man.* London: Collins, 1959.

Teller, Paul. "Relational Holism and Quantum Mechanics." *British Journal for the Philosophy of Science,* Vol. 37 (1986).

Thatcher, R. W., and John, E. R. *Functional Neuroscience,* Vol. I. New York: Lawrence Erlbaum Associates, 1977.

Thorpe, W. H. *Learning and Instinct in Animals.* London: Methuen and Co., 1963.

Thorpe, W. H. *Animal Nature and Human Nature.* London: Methuen and Co., 1974.

Todes, Samuel J. *The Human Body as the Material Subject of the World.* New York: Garland Publishing, 1989.

Treisman, Ann. "Features and Objects in Visual Processing." *Scientific American,* Vol. 255, No. 5 (November 1986).

Usdin, Gene, ed. *Overview of the Psychotherapies.* New York: Bruner/Mazel, 1975.

Wald, R. M. "Correlations and Causality in Quantum Field Theory." In R. Penrose and C. J. Isham, eds., *Quantum Concepts in Space and Time.* Oxford: Oxford University Press, 1986.

Walker, Evan Harris. "The Nature of Consciousness." *Mathematical Biosciences.* Vol. 7 (1970).

Watson, J. B. "Psychology as a Behaviourist Views It." *Psychological Review,* Vol. 20 (1913).

Weizenbaum, Joseph. *Computer Power and Human Reason.* London and New York: Penguin Books, 1984.

Wheeler, John Archibald. "Beyond the Black Hole." In Harry Woolf, ed., *Some Strangeness in the Proportion.* Reading, Mass.: Addison-Wesley Publishing Co., 1980.

Wheeler, John Archibald, and Zurek, Wojcieck Hubert, eds. *Quantum Theory and Measurement.* Princeton, N.J.: Princeton University Press, 1983.

Whitehead, A. N. *Process and Reality.* New York and London: Free Press, 1979.

Wigner, Eugene P. "Remarks on the Mind-Body Question." In John Archibald Wheeler and Wojcieck Hubert Zurek, eds., *Quantum Theory and Measurement.* Princeton, N.J.: Princeton University Press, 1983.

Wilber, Ken, ed. *The Holographic Paradigm and Other Paradoxes.* Boulder, Colo., and London: New Science Library, Shambala Press, 1982.

Wolf, Fred Alan. *Taking the Quantum Leap.* New York: Harper & Row, 1988.

Wordsworth, William. *The Prelude.* London and New York: Penguin Books, 1988.

Zeh, H. D. "On the Interpretation of Measurement in Quantum Theory." In John Archibald Wheeler and Wojcieck Hubert Zurek, eds., *Quantum Theory and Measurement.* Princeton, N.J.: Princeton University Press, 1983.

Zohar, Danah. *Through the Time Barrier.* London: Heinemann, 1982.

Zukav, Gary. *The Dancing Wu Li Masters.* London: Rider/Hutchinson, 1979.

Index

action-at-a-distance, *see* nonlocality
Adler, Alfred, 157
adolescence, quantum theory and, 9–10
aesthetics, 203–215
  boredom and, 208–209, 213–214
  functional needs and, 205, 206
  of Greeks, 207
  Newtonian physics and, 207
  overcrowding and, 209, 210
  of parks, 213–215
  quantum, 208–215
  relational holism and, 102–103
  of Romans, 207
  sculpting effect and, 210–211
  see also art; beauty
aggression, 160, 161
aging process, 21
alienation, 51–52, 131, 215, 219, 227
  architecture and, 206
  getting beyond, 17
  holism vs., 72, 74
  lack of world view and, 232, 233
  narcissism and, 155, 158, 159, 235
  overcrowding and, 210
  roots of, 17–20, 37
alpha waves, 89
amoebas, 54, 65, 66, 190
anesthetics:
  consciousness and, 54, 87
  self and, 166
animals, 181
  consciousness of, 53–55, 86
  Freud's views on, 160
  see also specific animals
animism, 55
Anthropic Principle, 224–225
antirealist view (Copenhagen Interpretation)
    of quantum theory, 22, 43
architecture, 18, 206–210
  boring, 208–209
  confusing, 209, 210
  functionalist, 208–209, 211
Aristotle, 173, 207, 218, 224
art, 168
  folk, 215
  pain and, 209
  painting as, 102–103, 164, 212
  relational holism of, 103–104, 164
astral body (shadow man), 143–144
atomism, 94
atoms, 21, 26
  Bohr, 30

atoms *(cont'd)*
  in covalent bonds, 134
  soul, 63
atoms of spirit, 103
attention:
  focused, 67, 69, 74, 76, 178
  selective, 116–117
autonomy vs. shame and doubt, 135, 136
awareness, "now" and, 120

bats, consciousness of, 53
beauty, 159, 161, 164, 167
  see also aesthetics
behavior:
  causes of, 181–182
  creative motivation for, 189
  meaning of, 182
  voluntary vs. involuntary, 171–176
behaviorism, 52, 235
being:
  contextualism and, 47
  doing equated with, 63–65
  functionalist view of, 63–65
  Heidegger's views on, 102–103
  oneness of, 37
  quantum theory and, 25–29, 34, 38, 47
  as wave/particle dualism, 25–29, 34
Being and Nothingness (Sartre), 130
Being and Time (Heidegger), 129–130, 161
Bell, John, 36
Bell's Theorem, 36
Bergson, Henri, 188
Bernstein, Leonard, 175
beta waves, 89
Big Bang, 225, 226
billiard balls, Newtonian, relationship of, 99, 129, 131, 132
biology:
  Darwinian, 219
  Freudian psychology and, 160
blood, 63, 86
Bloom, Allan, 155, 158, 159
boatmen analogy, nonlocality and, 36–37, 99–100
body, human:
  consciousness and, 62–63
  needs of, 204–205
  soul vs., 92–93, 108
  see also brain; mind/body problem
Bohm, David, 32, 33, 35*n*, 52, 96, 196, 230
  holographic paradigm and, 73, 74
  panpsychism and, 58–61
  thought processes compared with quantum processes by, 76–79
Bohr, Niels, 22, 27–28, 30
boredom:
  aesthetics and, 208–209, 213–214
  relieving, 189, 196
Bose-Einstein condensate, 83–87, 98, 100–102, 191, 221
  aesthetics and, 209
  bosons in, 104–105
  creativity and, 195
  freedom and, 176, 177, 179–180
  quantum identity and, 114–116, 120, 176
  quantum intimacy and, 133, 135–136, 139
  quantum psychology and, 163
  quantum vacuum and, 226
bosons, 104–106, 163, 194, 221–224, 236
  creativity and, 222–224
  quantum vacuum and, 227
Bragg, Sir William, 26
brain:
  Bose-Einstein condensates and, 85–87, 177, 179–180, 191, 195, 221
  cerebral cortex, 66, 72, 90, 109
  computer model of, 62–70, 72, 74, 78, 87, 176
  consciousness and, 62–75, 78, 85–89, 109
  damage to, 63, 66, 87
  drugs and, 63, 66, 87
  evolution of, 65–66
  focused attention and, 67, 69, 74
  forebrain, 66
  holographic model of, 70–75, 78
  indeterminacy and, 79, 80
  mind vs., 101–102
  neurons in, 64, 67, 79, 82, 85–86, 87, 89–90, 108, 115, 150, 176, 180, 221
  physical functions of, 63–64
  pineal gland of, 63
  as quantum computer, 179–180
  right vs. left hemisphere of, 72, 109
  schizophrenia and, 116*n*
  structure of, 10, 108
  unity of person and, 108–110
brain stem, 109
breast:
  as material world, 204–205
  as source of consciousness, 63
Browning, Elizabeth Barrett, 137
Browning, Robert, 137
Buber, Martin, 102, 128–129, 130, 216
Buddhism, 107, 110–111
  "science" of, 217–218
bunching effect, photon, 222–223

Cahoone, Lawrence, 199, 234
Calder, Nigel, 57*n*

cancer, 21
Capra, Fritjof, 48, 111
Cartesian philosophy, 18, 43, 51–52, 63, 92–93, 144, 236
  intimacy and, 129–131
  psychology and, 156
cathexis, 160, 163
cats, consciousness of, 53
cause-and-effect relation, 24–25
cells:
  microwave radiation and, 84–85
  yeast, 84, 86, 189
  see also neurons
cerebral cortex, 66, 72, 90, 109
chance, 31
Chaos (Gleick), 177
charisma, 116
chess, as relational holism, 102
children:
  creativity of, 189–194
  mother's relationship with, *see* mother-child relationship
  mother's separation from, 146
  psychic development of, 133–136
  responsibility and, 181, 184–185
choice, 178–185
  concentration and, 178–180
  creativity and, 196–198, 200–202
  meaning of, 200, 201
  probability of, 183–185
  rationality and, 181
  reasons for, 184
Christianity, 18, 93, 201, 216–218, 229
  aesthetics of, 207
  consciousness and, 51, 52
  determinism of, 173
  existentialism and, 144–145
  science of, 218
classical physics, *see* Newtonian physics
clockwork universe, 18, 28
Closing of the American Mind, The (Bloom), 155
coauthored world, 198–199
coherence:
  evolution of, 227, 229
  of personality, 116
  of world view, 232–234, 236
coherent biphotons, 85
coherent states, *see* Bose-Einstein condensate
coins, relational holism and, 99
collective unconscious, 158
commitment, 155, 159–168
  basis for, 165, 166
  existentialist emphasis on, 161–162
  Freudian psychology and, 160, 161
  motivation for, 165
  quantum view of, 164–168
  transforming power of, 167
  transpersonal vs. interpersonal, 167
  values and, 159–160, 166–167
communitarianism, 235, 237
compasses, in condensed phase, 81–82
Complementarity, Principle of, 26, 28–29
computer model of brain, 63–70, 87
  cost of, 72
  determinism and, 176
  holographic paradigm compared with, 72, 74
  inadequacies of, 67–70, 78
computers, 189
  parallel processing of, 64, 72
  quantum, 105, 179–180
concentration, mental, 76, 177–180
condensed phases, 81–87, 90–91
  compasses in, 81–82
  pumped system and, 83–87, 90
conflict:
  in Freudian psychology, 160–161
  personality effects of, 116
confusion, aesthetics and, 208, 209–210
consciousness, 10–11, 28, 31, 50–91
  of animals, 53–55, 86
  borderline areas of, 177–178
  brain and, 62–75, 78, 85–91, 109
  Buddhist conception of universe and, 218
  Cartesian view of, 43, 51–52
  computer model of brain and, 63–70
  evolution of, 220–225, 230
  form of, 66
  genealogy of, 220–225
  Greek view of, 63
  holographic paradigm and, 70–75
  human needs and, 205–206
  "islands" of, 83–84
  Judeo-Christian tradition and, 51
  matter and, 10, 23, 43, 56, 62–75, 90–91, 220–222
  as missing link, 42, 43
  Newtonian physics and, 18, 67, 69–70, 81, 219
  nonlocality and, 37, 78–79
  panpsychism and, 55–61
  of particles, 59–61
  physics of, 42
  Popper's views on, 58*n*
  primitive, 104
  as problem, 22
  quantum vacuum compared with, 227
  Schrödinger's cat and, 38–51
  selves within, 112–113
  steady state of, 81

consciousness *(cont'd)*
unity of, 69–70, 73–75, 77, 80–81, 100–104, 108, 115–116
use of term, 220
consciousness, quantum mechanical model of, 11, 17, 23, 37, 43, 54*n*, 76–91, 219, 236
aesthetics and, 207–208
Bohm's analogies and, 76–79
Bose-Einstein condensates and, 83–87
brain's two interacting systems in, 90
condensed phases and, 81–87, 90–91, 100
freedom and, 177, 180
injury and, 169
Marshall and, 79–80
Penrose and, 78–80
quantum psychology and, 159, 163
relationship and, 100–104
contextualism, 47–48
contingent truth, 129
Copenhagen Interpretation (antirealist view) of quantum theory, 22, 43
Copernican revolution, 18
Copernicus, Nicolaus, 207
corpus callosum, 109
correlation experiments, 99–100
photon, 36–37
twin, 35–36, 79
covalent bonds, 105, 224*n*
psychic development compared with, 134
creative fidelity, 144–145
creativity, 64, 72, 186–202, 236
bosons and, 222–224
of children, 189, 191
coauthored world and, 198–199
discovery and, 192–194, 197, 204–205
of earthworms, 194–195
evolution and, 194
growth and, 189
habits vs., 186
indeterminism and, 80
of life, 189, 221–222
living order and, 189–190
material world and, 203–204
morality and, 195–201
quantum impulse and, 193–194
quantum memory and, 123–124
as *raison d'être,* 188
responsibility and, 231
criminals, responsibility of, 171–172, 176
Critical Dictionary of Psychoanalysis (Rycroft), 130, 160
Crook, John, 81
Culture of Narcissism, The (Lasch), 154, 155

Dada movement, 209
Darrow, Clarence, 175
Darwin, Charles, 18
Darwinian biology, 219
death, 10
creative fidelity and, 144–145
life after, *see* immortality; immortality, quantum
of Schrödinger's cat, 39–43
defense mechanisms, 159
delayed-choice experiment, 45, 47, 59, 197–198
delta waves, 89
Descartes, René, 20, 51, 63, 70, 87, 92–93, 95, 112
"I" of, 129, 131
determinism, 219
behavior and, 173–176
Christian view of, 173
existentialist view of, 162
free will, 10, 80, 173–176
of Freudian psychology, 160, 174–175
Greek view of, 173
historical, 174, 175
in Newtonian physics, 27, 174
scientific, 27, 173–176
Diamond Sutra, 73
dipoles, in pumped system, 83, 84
DNA, 194, 221
photons in, 85
DOCTOR, 64
dogs, consciousness of, 53
doing, being equated with, 63–65
Donne, John, 168
doubt and shame vs. autonomy, 135, 136
dreams, 159
lucid, 115–116
drugs, 158
brain and, 63, 66, 87
dualism, 62
Cartesian, 43, 51, 63, 70, 92–93, 144, 236
immortality and, 144
Platonic, 144
wave/particle, *see* wave/particle duality
see also mind/body problem
Duino Elegies, The (Rilke), 188

earthworms, 66, 190
creativity of, 194–195
EEG patterns, 87, 89–90
ego, 130–131, 160–161, 169
egocentricity, 18, 159
ego instinct, 126

Einstein, Albert, 24, 28
  nonlocality and, 35–36
  relativity theory of, 20, 24
  Theory of Hidden Variables of, 35–36
Einstein, Podolsky, Rosen (E.P.R.) Paradox, 35
electrons, 23, 26–28, 79, 104, 134, 224
  "berserk," 209
  in Bohr atoms, 30
  consciousness of, 59–61
  fluctuating existence of, 117–118
  measuring conundrum of, 26–27, 41–42, 76
  as probability wave, 31–32
  quantum leaps of, 30, 31
  relationship of, 99
Eliot, T. S., 38, 124, 141
ELIZA, 64
empathy, 10, 127
Encyclopedia of Philosophy, 56, 175
energy:
  concentration and, 180
  habit and, 185–186
  probability and, 183
England, 175, 207
entropy (Second Law of Thermodynamics), 85, 190
Epicurus, 63
epilepsy, split-brain surgery and, 109
epistemology, 47, 48
E.P.R. (Einstein, Podolsky, Rosen) Paradox, 35
Erikson, Erik, 134–136
essence of quantum self, 162–163, 184
evolution, 18, 21, 219
  of brain, 65–66
  of coherence, 227, 229
  of consciousness, 220–225, 230
  increasing rhythms of, 225
  responsive, 194
  virtual transitions and, 33
evolutionary paradigm, 200, 221
"Excitations of a Bose-Einstein Condensate" (Marshall), 180*n*
existence, 98, 111, 112
  of electrons, 117–118
  fluctuating, of particles, 117–118
  of love, 164
  meaning of, 219
  of quantum self, 162–163
existential hero, 19
existentialism, 110, 129–130, 161–164
  Christian, 144–145
  freedom and, 182–183
  morality and, 168
existential psychoanalysis, 157
experience:
  consciousness and, 52
  of the dead, 145
  of freedom, 172
  holographic paradigm and, 72–75
  Ideas vs., 18
  narcissism and, 154
  unity of, 69–70, 73–75

fact gathering vs. rapport, 27
faith, 217, 226
  leap of, 182
Fall, the, 227, 229
fantasy, 130–131, 138
fate, 173
Fechner, G. T., 56
Feigl, Herbert, 94, 95, 97
femininity, 151–153
fermions, 104–106, 114*n,* 224, 236
fidelity vs. infidelity, 196–198, 200, 201
film, quantum motion compared with, 30
Finkelstein, David, 225
Finnegans Wake (Joyce), 209
Flies, The (Sartre), 182–183
focused attention, 67, 69, 74, 76, 178
forebrain, 66
Four Quartets (Eliot), 141
Fran, Jerome, 157–158
free enterprise, 175
free will, freedom:
  choice and, 178–185
  Christian view of, 173
  consciousness and, 54
  creativity and, 196–202
  determinism vs., 10, 80, 173–176
  existential view of, 162, 163, 182–183, 199
  physical basis for, 177
  quantum indeterminacy and, 177–180
  responsibility and, 171–187
Freud, Sigmund, 18, 126, 130, 157–162, 164, 182, 235
  determinism of, 160, 174–175
  morality as viewed by, 169
  oceanic feeling of, 111, 142
  on primary process, 178
Fröhlich, Herbert, 83–87, 90, 190, 193
Fröhlich-style Prigogine systems, 193*n,* 208, 221
functionalism, 63–65
  boredom and, 208–209
functional needs, 205, 206

Gaia hypothesis, 56
Galileo, 207
gardens, Japanese, 210, 211
genetic mistakes, 21
"Gestalt Prayer," 155
Gestalt therapy, 113*n*, 155
Gleick, James, 177
God:
  Christian, 218, 233
  immortality and, 143
  in Judeo-Christian tradition, 233
  man's relationship with, 51
  new physics and, 226–227, 229, 230
  world view and, 233
golden mean, 207
Golden Rule, 199–200
Grace, 229
Graves, Robert, 125–127
gravity, 223
Great Britain:
  free enterprise and competition in, 175
  Victorian buildings in, 207
Great Chain of Being, 55–56, 73
Great Chain of Being (Lovejoy), 111
Greeks, ancient, 18, 23, 63
  aesthetics of, 207
  fate of, 173
  religion-science link of, 217
group mind phenomenon, 139
group therapy, 157
guilt vs. initiative, 136

habit, 185–186
Hawking, Stephen, 230
Hebrew legends, 45, 47
Heidegger, Martin, 102–103, 110, 123*n*, 161, 164
  aesthetics of, 211, 212
  on interpersonal relations, 129–130
Heisenberg, Werner K., 48
  quantum indeterminism as viewed by, 27–28
  Uncertainty Principle of, 21, 26–31, 45, 76, 209
helium-4, 105
Heraclitus, 55, 96
heroes, 186
  existential, 19
  tragic, 173
Hey, Tony, 225
Hidden Variables, Theory of, 35–36
Hill, Carol, 57*n*
Hindu philosophy, 63
historical determinism, 174, 175
history, 18, 19, 20, 126
holism, 72–75, 87
  relational, *see* relational holism
holograms, 87
  construction of, 70, 74
  definition of, 70
holographic paradigm, 70–75
  computer model compared with, 72, 74
  inadequacies of, 73–75, 78
Housman, A. E., 107
humanistic psychology, 157
Humanizing the Narcissistic Style (Johnson), 156*n*
human needs, 205–206
hussy, quantum-level, 32–33, 80, 84, 178, 196

"I," 62, 74, 102, 107–108
  in Cartesian philosophy, 129, 131
  inability to reduce, 114–115
  "not-I" vs., 125–131
  see also identity, personal
I and Thou (Buber), 216
I-centered culture, 18, 154–155, 158
id, 160–161, 169, 209
idealism, 95–96, 234–235
Ideas vs. experience, 18
identical-twins analogy, nonlocality and, 35–36
identification:
  with children, 146–147
  projective, 127, 130–131, 133–134, 137, 138, 142, 166
identity, personal, 10, 106–124
  brain and, 108–110, 116*n*, 118–119
  Buddhist view of, 107, 110–111
  conflict and, 116
  memory and, 118–124
  Parfit's denial of, 110–111
  particles compared with, 113–114, 117–118
  quantum, 106, 112–124
  relational holism and, 113–114
  relationships and, *see* intimacy, quantum
  responsibility and, 176
  Schrödinger's cat and, 38–41
  selves within, 112–117, 126
  sickness and, 116
  sleep and, 115–116
  split-brain research and, 108–112
  unity of, 108–116, 133
I-It, 128, 130
imagination, 64, 72, 109

immortality:
in Jungian psychology, 158
traditional views of, 143–144
immortality, quantum, 140–153
motherhood and, 141–142, 146–147, 152
particles and, 142–143
quantum intimacy and, 145, 147–150
quantum memory and, 145–150, 152
surrender and, 152
impulse, quantum, 193–194
indeterminism, quantum, 27–28, 34
brain and, 79, 80
freedom and, 177–180
individualism, 235, 237
infants, breastfeeding of, 204–205
infidelity vs. fidelity, 196–198, 200, 201
information, holographic paradigm and, 70, 72
initiative vs. guilt, 136
insects, nervous system of, 65
intention, 80
intimacy, 124–140
Cartesian philosophy and, 129–131
empathy and, 127
fear of, 154
I-Thou and, 128–131
love and, 125
in mother-child relationship, 126, 130–131, 133–136
Newtonian physics and, 129–131
as object representation, 130, 131
problem of, 129–131
projective identification and, 127, 130–131, 133–134, 137, 138
in psychotherapy, 126–127, 138
intimacy, quantum, 131–140
Bose-Einstein condensate and, 133, 135–136, 139
commitment and, 165–168
covalent bonding compared with, 134
immortality and, 145, 147–150
psychic development and, 133–137
quantum memory and, 137, 147
quantum psychology and, 165–169
role reversals and, 138
wave/particle duality and, 131–132, 137, 139–140
intuition, 72, 109, 172
isolation:
Cartesian-Newtonian paradigm of, 129–131
narcissism and, 155, 156
poverty of material world and, 213
quantum psychology and, 169
I-Thou, 102, 128–131

James, William, 56, 120
Japanese gardens, 210, 211
jellyfish, nervous system of, 65, 66
Jennings, H. S., 54
Johnson, Stephen, 156*n*
Josephson, Brian, 179
Josephson Junctions, 179
Joyce, James, 209
Judaism, 201, 229
mysticism and, 188
Judeo-Christian tradition, 51, 218
world view of, 233–234
Jung, Carl Gustav, 111, 158, 170, 198, 233
on Pueblo Indians, 229–230
Jungian psychology, 127, 158, 159

Kant, Immanuel, 199
Kennedy, John F., 148
Kierkegaard, Søren A., 182
Klein, Melanie, 130–131, 193, 204
knowledge:
without context, 219
shared subjectivity and, 199
Krishnamurti, 10, 170

language, quantum theory and, 21
Lasch, Christopher, 154, 155, 162
lasers, 81, 84, 133
law, 171–172, 176
beauty and, 207
Lawrence, D. H., 137
Le Corbusier, 18, 206
Leibnitz, Gottfried Wilhelm von, 56
light, particles of, *see* photons
light waves, 26
holograms and, 70, 74
Lincoln, Abraham, 148
liver, consciousness arising from, 63
living order, 189–190, 221, 227
living system, nonliving system compared with, 223
logical thinking, 72, 77, 109, 181
loneliness, 158
Lotze, Rudolf Hermann, 56
love, 155, 161, 164, 167, 169
intimacy and, 126
as relational holism, 102
Lovejoy, Arthur, 111
Lovelock, J. E., 56
lucid dreams, 115–116
Lucretius, 103
Luther, Martin, 173

McGinn, Colin, 97
machine model, *see* computer model of brain
"machines for living," 206
mammals, forebrain of, 66
Many-Worlds Theory, 33
Marcel, Gabriel, 144–145, 148
marriage:
  fidelity vs. infidelity in, 196–198, 200, 201
  stability of, 166
Marshall, I. N., 10, 79–80, 142, 180*n*
Marx, Karl, 18, 174, 175
Marxism, 126, 235, 237
materialism, 51, 94–95, 101, 103, 235
matter, material world:
  aesthetics and, 203–215
  Christian view of, 93, 218
  conscious dealings with, 203–204, 206, 212
  consciousness and, 10, 23, 43, 56, 62–75, 90–91, 220–222
  materialist view of, 94–95
  mind vs., 43, 51, 62, 70, 91
  Newtonian physics and, 20, 24, 93–94
  panpsychism and, 55–61
  Platonic view of, 93
  poverty of, 211–214
  proto-mental properties and, 58
matter-consciousness relationship, 17
matter waves, 34
measurement, 28
  of wave function, 41–42, 45
  wave/particle duality and, 26–27, 41–42
mechanical model of universe, 18
medical psychiatry, 158, 159
medieval period, 56
Meditations (Descartes), 92
memory, 94
  Bose-Einstein condensates and, 85
  identity and, 118–124
  reliability of, 118–119
memory, quantum, 119–124
  in psychotherapy, 123, 136, 147
  quantum immortality and, 145–150, 152
  quantum intimacy and, 136, 137
mentally handicapped, responsibility and, 184–185
Merleau-Ponty, Maurice, 47
Mies Van Der Rohe, Ludwig, 210, 211
Miller, Arthur, 168
mind:
  brain vs., 101–102
  creation and, 224
  of God, 230
  group, 139
  matter vs., 43, 51, 62, 70, 90
mind/body problem, 62, 91–106
  Cartesian view of, 43, 51, 70, 92–93, 95
  idealist view of, 95–96
  materialist view of, 94–95
  panpsychism and, 96–97, 103
  quantum-level matter and, 98–106
  wave/particle duality and, 98, 100–101, 103
  world view and, 234–237
missing links:
  consciousness as, 42, 43
  mutations and, 33
mistrust vs. trust, 134, 136
modernism, 19, 235
morality, 168–170
  creativity and, 195–201
  freedom and, 186
  Golden Rule and, 199–200
  Sartre's views on, 168, 198–199
  see also responsibility
moral relativism, 200
mother-child relationship, 126, 130–131, 133–136, 166
  quantum memory and, 146–147
  separations and, 146
motherhood:
  modern physics and, 9, 10
  quantum immortality and, 141–142, 146–147, 152
motion, *see* movement
motivation:
  for commitment, 165
  creative urge, 189
movement:
  in Newtonian physics, 18, 24–25, 29–30
  in quantum theory, 29–34
music, quantum intimacy compared with, 137
mutations, 33
mysticism, 229, 235
  Eastern, 74–75, 111, 126, 237
  Jewish, 188

Nagel, Thomas, 53, 57–59, 90–91, 94
  on unity of person, 110
narcissism, 154–162, 165, 235
  alienation and, 155, 158, 159, 235
  as disease of relationship, 155, 165
  psychological model of, 155–162
  self-hatred and, 155
  sources of, 156

symptoms of, 155, 156
three expressions of, 156
Nature, 28
animist, 55
commitment to, 165, 167, 168
consciousness and, 43
culture vs., 235, 237
intellect and, 23
Newtonian physics and, 20, 207
virtual transitions in, 33
needs:
aesthetic, 206–215
bodily, 204–205
functional, 205, 206
human, 205–206
neo-Darwinism, 219
Neoplatonism, 93
nervous system, 63–66
brain and, 63–64
evolution of, 65–66
Neuman, Erich, 111*n*
neurobiology, 78
neurons:
in brain, 64, 67, 79, 82, 85–86, 87, 89–90, 108, 115, 150, 176, 180, 221
firing of, 79, 85–86, 89, 176, 180, 221
neutrons, 104, 105
new physics, *see* quantum theory
Newtonian (classical) physics, 18–20, 22, 24–25, 52, 77, 112, 236
aesthetics and, 207
brain and, 67, 69–70
consciousness and, 18, 67, 69–70, 81, 219
determinism of, 27, 174
holograms and, 74
intimacy and, 129–131
matter and, 20, 24, 93–94
movement in, 24–25, 29–30
Nature and, 20, 207
problems of defining freedom in, 177
psychology and, 156
quantum theory compared with, 26, 28, 29–30
"new towns," 209
Nicolis, Gregoire, 223
Nietzsche, Friedrich W., 123*n*
nihilism, 110
nonlocality (action-at-a-distance), 34–37, 78, 149, 169
consciousness and, 37, 78–79
correlation experiments and, 35–37, 99–100
intimacy and, 133
in waves vs. particles, 37
"now," 120, 123
now-centered culture, 154

objectivity, 48, 219, 234
object relations, 157, 160
object representation, 130, 131
observation, observer, 41–50, 225
in delayed-choice experiment, 45, 47
reality and, 41–49
reality "created" by, 48
shared subjectivity and, 199
type of, 44–49
observer-participancy, 45, 47–48
oceanic feeling, 111, 142
"Ode" (O'Shaughnessy), 231
oneness of being, 37
optic nerve, 21
Order out of Chaos (Prigogine and Stengers), 203
Orlov, Yuri, 80
O'Shaughnessy, Arthur William Edgar, 231
overcrowding, aesthetics and, 209, 210

painting, 164, 212
as relational holism, 102–103
panpsychism, 55–61, 90
history of, 55–56
limited, 56–61, 96–97
mind/body problem and, 96–97, 103
parallel processing, 64, 72
paramecium, 65
parents:
identification with children by, 146–147
see also mother-child relationship; motherhood
Parfit, Derek, 110–111, 126
memory as viewed by, 118–119, 120, 124
parks, 213–215
Parmenides, 55, 96
Participatory Anthropic Principle, 225
particles:
consciousness of, 59–61
death and, 142–143
light, *see* photons
in Newtonian physics, 26
nonlocality and, 37
primitive consciousness and, 104
in quantum theory, 25–31, 37, 104–106
quantum vacuum and, 225
of relationship, 104–106, 163, 224

particles *(cont'd)*
self compared with, 113–114, 117–118
types of, 104–106
see also wave/particle duality
past:
denial of, 162
quantum memory and, 122–124, 136, 145–148
patient-therapist relationship, 27, 157
patterns, continuity of, 143
Paul, Saint, 93, 196
Penrose, Roger, 78–79, 80
perception:
Descartes's views on, 87
development of, 134–135
visual, 67, 69, 74
peristalsis, 65
Perls, Fritz, 113*n*
Phaedo (Plato), 93
phases:
condensed, 81–87, 90–91
defined, 81
phenomenology, 47, 67
Phenomenon of Man, The (Teilhard de Chardin), 171
philosophy:
Cartesian, 18, 43, 51–52, 63, 92–93, 129–131, 144, 156, 236
existential, 110, 129–130, 144–145, 161–164, 168, 182–183
individual vs. relationships in, 126
photon bunching effect, 222–223
photons, 21, 23, 41–42, 104
Bose-Einstein condensate and, 85, 87, 221
brain neurons and, 79
coherent ordering of, 85, 90, 98
correlation experiments and, 36–37
in delayed-choice experiment, 45, 47, 59, 197–198
in living vs. nonliving systems, 223
in two-slit experiment, 45, 59
wave/particle duality and, 45, 59
physics:
classical, *see* Newtonian physics
of everyday life, 10, 17–23
relativity theory in, 20, 34
as world apart, 10
see also quantum theory
pineal gland, 63
Pippard, Brian, 217
Planck, Max, 30
plants, 86
panpsychism and, 55, 56
plastic, 211–212
Plato, 18, 56, 72, 102, 144, 164
on beauty, 207
Republic of, 93, 167
Platonism, 93, 218
pleasure principle, 157
Podolsky, Boris, 35*n*
poetry, 177–178, 188
quantum systems compared with, 211
politics, determinism and, 176
Popp, Fritz, 85, 90, 194, 222–223
Popper, Karl, 58
possibilities, 31–33
Schrödinger's cat and, 39–40
predictions, of quantum theory, 21–22, 27–28
present, quantum memory and, 122–124, 136, 145–148
Prigogine, Ilya, 47, 190, 194, 199, 203, 225
evolutionary paradigm of, 200, 221
quantum systems of, 193*n*, 208, 221, 223
primary process of mental functioning, 178
Principle of Complementarity, 26, 28–29
probabilities, 27–28, 31–32
of choices, 183–185
of collapse of quantum wave function, 183
probability waves, 31–32, 34, 39
process:
holographic paradigm and, 73
life as, 151–152
process theology, 226–227
projective identification, 127, 130–131, 133–134, 137, 138, 142, 166
Protestantism, 173
proto-mental properties, 58–59
protons, 35*n*, 104, 105, 224
psychiatric interview, first, 27
psychiatry:
forebrain and, 66
medical, 158, 159
psychic development, 133–137, 159
Erikson's stages of, 133–136
psychoanalysis, 157–159, 182
determinism of, 160, 174–175
morality and, 169
passivity of analyst in, 157
relationship and, 126, 130–131
psychological processes, virtual transitions compared with, 33
psychology, 18, 19, 21, 78
behaviorist, 52
Jungian, 127, 158, 159
physics wedded to, 23
religion replaced by, 218

psychology, quantum, 159, 162–170
  commitment and, 164–168
  intimacy and, 165–169
  morality and, 168–170
psychotherapy, 113, 157–158, 182
  intimacy in, 126–127, 138
  quantum memory in, 123, 136, 147
  rebirth and, 201–202
  role reversals in, 138
  value-free, 169
Ptolemy, 218
Pueblo Indians, 229–230
pumped system, 83–87, 90
purpose of life, 22, 221–222, 225, 226, 233

quantum computers, 105, 179–180
quantum field theory, 29, 225
quantum identity, *see* identity, personal, quantum
quantum immortality, *see* immortality, quantum
quantum impulse, 193–194
quantum intimacy, *see* intimacy, quantum
quantum leap:
  of electrons, 30, 31
  nonscientific use of term, 21
quantum memory, *see* memory, quantum
quantum psychology, *see* psychology, quantum
quantum responsibility, 176–187
quantum superposition effects, 179, 211
quantum theory (quantum mechanics), 20–37
  antirealist view (Copenhagen Interpretation) of, 22, 43
  author's discovery of, 9–10
  being and, 25–29, 34, 38, 47
  Bohr-Heisenberg interpretation of, 27–28
  consciousness and, *see* consciousness, quantum mechanical model of
  incompleteness of, 21–22, 28, 43–44
  language and, 21
  Many-Worlds Theory of, 33
  as metaphor, 11, 21
  movement and, 29–34
  predictions of, 21–22, 27–28
  relationship and, 34–37; *see also* intimacy, quantum
Quantum Theory (Bohm), 76
quantum vacuum, *see* vacuum, quantum
quantum world view, 236–237

radiation, in cells, 84–85
rape, responsibility for, 171–172, 176
rapport vs. fact gathering, 27
rationality, 160–161, 172, 181
  creativity and, 195
reality:
  Bohm's views on, 58–59
  Bohr-Heisenberg view of, 27–28
  Cartesian view of, 51
  indeterminism of, 27–28
  Nagel's description of, 58
  Newtonian view of, 24–25
  observation and, 41–49
  quantum level of, 22, 27–28, 30–31, 38–39, 58, 164, 226
Realm of Forms, 207
real transitions, 32
rebirth, 201–202
relational holism, 99–106, 164, 202, 221
  identity and, 113–114
relationship, 34–37, 98–104
  commitment and, 159–163
  with the dead, 144–145
  external, 99
  holographic paradigm and, 73
  internal, 99
  interpersonal, *see* intimacy; intimacy, quantum
  matter-conscious, 17
  mother-child, *see* mother-child relationship
  narcissism as disease of, 155, 165
  particles of, 104–106, 163, 224
  patient-therapist, 27, 157
  quantum world view and, 237
  world view and, 232
relativism, 20, 200, 235
relativity theory, 20, 34
religion, 155, 156, 216–219
  creativity and, 188
  quantum vacuum and, 227, 229
  rebirth and, 201
  science vs., 217–218, 220
  see also Buddhism; Christianity; Judaism
Renaissance, 56
Republic (Plato), 93, 167
responsibility, 112, 133, 169–187
  of children, 181, 184–185
  creativity and, 231
  of criminals, 171–172, 176
  determinism and, 173–176
  quantum, 176–187
  quantum identity and, 176
reversibility, 31

Rilke, Rainer Maria, 188, 213, 231
rituals, value of, 123–124, 166–167, 229–230
Rogers, Carl, 157
role reversals, 138
Romans, ancient, aesthetics of, 207
Rosen, Nathan, 35*n*
Russell, Bertrand, 18–19, 174
Rycroft, Charles, 160, 174–175

sacrifice, 155, 161
Sartre, Jean-Paul, 110, 129, 130, 161–164
  on freedom, 182–183
  morality as viewed by, 168, 198–199
schizophrenia, 116*n*
Schrödinger, Erwin, 39
Schrödinger's cat, 22, 38–49, 84, 223
  consciousness of, 50–51
  freedom and, 177, 178
  observation problem and, 41–50
Schrödinger wave function, 39–40, 42–44, 48
science, 217–220
  determinism of, 27, 173–176
  mechanical, 219; *see also* Newtonian physics
  religion vs., 217–218, 220
scientific revolution, 18, 218
sea anemone, 54
Second Law of Thermodynamics (entropy), 85, 190
self:
  defined, 114
  illusion of, 107, 110–111, 162
self-fulfillment, 157, 158
separateness, notion of, 25, 34
  brain and, 67, 69
  escape from, 111
  in Freudian psychology, 157
  holograms and, 74
separation anxiety, 146
Serres, Michel, 20, 23
sex instinct, 126, 157, 160
shadow man (astral body), 143–144
shame and doubt vs. autonomy, 135, 136
"Shropshire Lad, A" (Housman), 107
sickness, self and, 116
sleep:
  dreams and, 115–116, 159
  EEG patterns in, 89
  self in, 115–116
smoking, giving up, 181–182, 184
snails, consciousness of, 54
Socrates, 93
soul, 87, 91
  body vs., 92–93, 108
  immortal, 143, 144
soul atoms, 63
space, 225
  correlation effects across, 34–36
  in Newtonian physics, 24–25, 26
  in quantum theory, 25, 30, 34–36, 39
  in relativity theory, 20
  Schrödinger's cat and, 39
specious present, 120
spine, 63, 66
Spinoza, Baruch, 56, 58, 73, 96
spiritualism, 143
split-brain research, 108–112, 144
steady state, 81
Stengers, Isabelle, 193*n*
Stuart, C.I.J.M., 85*n*
subatomic particles, *see* particles
subjectivism, 48, 52, 96, 234, 235
  shared, 199
sublimation, 161, 164
Sun, sons of, 229–230
superconductors, 81, 84
  brain as, 83
  covalent bonds of, 105
superego, 161, 169
superfluids, 81, 84, 105
  brain as, 83
surgery, split-brain, 108–109
surrender, 152
Symposium (Plato), 102

Taylor, Charles, 183, 197, 202
technology, 18, 52, 219, 235
  determinism and, 174
Teilhard de Chardin, Pierre, 56, 171, 226–227
teleology, Aristotelian, 218, 224
telepathy, 10
temperature, body, condensed phases and, 83, 84
Thatcher, Margaret, 175
theology, process, 226–227
Theory of Hidden Variables, 35–36
therapist-patient relationship, 27, 157
Thermodynamics, Second Law of (entropy), 85, 190
theta waves, 89
"Thieves, The" (Graves), 125–127
Thorpe, W. H., 54
Thou, 102

thought:
  freedom of, 80
  "I" and, 129
  logical, 72, 77, 109, 181
  magical, 178
  quantum processes compared with, 76–79
  rational, 160–161, 172, 181, 195
  unity of, 77
  see also mind
thought experiments:
  E.P.R. Paradox and, 35*n*
  see also Schrödinger's cat
Timaeus (Plato), 56, 102
time, 225
  correlation effects across, 36–37
  dream, 111–112
  in Newtonian physics, 24, 26
  "now," 120, 123
  in quantum theory, 30, 34, 36–37, 39
  in relativity theory, 20
  Schrödinger's cat and, 39
Todes, Samuel, 192*n*
tragedy, Greek, 173
trust vs. mistrust, 134, 136
truth, 159, 164, 167
  contingent, 129
  Heidegger's views on, 102–103
  lived, 232
  narcissism and, 154
  perspective vs., 48
  relational holism of, 102–103
"truth within a situation," 47–48
twins analogy, nonlocality and, 35–36, 79
two-slit experiment, wave/particle duality and, 45, 59

Uncertainty Principle, 21, 26–31, 45, 76, 209
unconscious, 160–161
  collective, 158
  determinism and, 174–175
unity:
  condensed phases and, 81–87, 90–91
  of consciousness, 69–70, 73–75, 77, 80–81, 100–104, 108, 115–116
  of experience, 69–70
  in Jungian psychology, 158
  of person, 108–116, 133
  of quantum systems, 77
  relational holism and, 99–106
universe:
  Buddhist conception of, 217–218
  clockwork, 18, 28
  consciousness and, 22, 43
  consciousness compared with, 101
  Greek and medieval, 18
  order of, 223
  participatory, 45
  quantum vacuum and, 225–226
urban life:
  material poverty of, 209, 210, 212–214
  parks and, 213–215

vacuum, quantum, 142, 164, 225–230
  naming of, 225
  religion and, 227, 229
value judgments, 48
values:
  articulation of, 197–199
  commitment and, 159–160, 166–167
  creativity and, 197–199
  existential view of, 162
  quantum view of, 164
  sublimation and, 161
van Gogh, Vincent, 102–103, 164, 212
violence, 19
virtual transitions, 32–33, 34, 80, 196, 230
vision (perception), brain and, 67, 69, 74
vision, moment of, 123*n*

Walters, Patrick, 225
Washington, George, 148
wave functions, waves:
  collapse of, probability of, 183
  concentration and, 178–179
  intimacy and, 132–140
  matter, 34
  in Newtonian physics, 26
  nonlocality and, 37
  overlapping, 132–133, 137, 147
  primitive consciousness and, 104
  probability, 31–32, 34, 39
  quantum memory and, 120, 122–123
  Schrödinger, 39–40, 42–44, 48
wave packet, 27
wave/particle duality, 25–29, 34, 58, 113
  being as, 25–29, 34
  focused attention vs. musing compared with, 76
  holographic paradigm and, 73
  measuring conundrum and, 26–27, 41–42
  mind/body problem and, 98, 100–101, 103
  observation and, 41–42

wave/particle duality *(cont'd)*
  quantum intimacy and, 131–132, 137, 139–140
  Schrödinger's cat and, 39, 41–42
  split brain compared with, 72
  two-slit experiment and, 4, 59
waves of spirit, 103
West Side Story (Bernstein), 175
"What Is It Like to Be a Bat?" (Nagel), 53
Wheeler, John Archibald, 43, 45, 47, 59, 197–198
  Participatory Anthropic Principle of, 225
Whitehead, Alfred North, 52, 56, 58, 66, 96
Wigner, Eugene, 43
Wilber, Ken, 70
wisdom, 72
withdrawal reflex, 65–66
Wordsworth, William, 193
work, dehumanizing, 189
worlding, 211
world view:
  functions of, 232–233
  Judeo-Christian, 233–234
  mechanical, 234–235
  quantum, 236–237

yeast cells, 84, 86, 189
Yeats, William Butler, 234
yoga meditation, 63

Zeh, H. D., 139
Zukav, Gary, 111

A Note About the Author

Danah Zohar was born in the United States. She received her bachelor's degree in physics and philosophy from the Massachusetts Institute of Technology and completed three years of postgraduate study in philosophy and religion at Harvard University, where she was a student of Erik Erikson. She is married to Dr. I. N. Marshall, a psychiatrist and psychotherapist, with whom she collaborated on this book, and they live in Oxford, England, with their two young children. *The Quantum Self* is part of a trilogy that will include the forthcoming volumes *The Quantum Society* and *The Quantum Spirit.*